Collecting
Volunteer Militaria

BY THE SAME AUTHOR

The Austin Seven, 1922–1939
Cars
Lord Austin—the man

Collecting
Volunteer Militaria

R. J. Wyatt

David & Charles
Newton Abbot

0 7153 6296 8

© R. J. Wyatt, 1974

All rights reserved. No part of this publication may be reproduced, stored in a retrieval system, or transmitted, in any form or by any means, electronic, mechanical, photocopying, recording or otherwise, without the prior permission of David & Charles (Holdings) Limited

Set in 11 on 13 point Times
and printed in Great Britain
by Latimer Trend & Company Ltd Plymouth
for David & Charles (Holdings) Limited
South Devon House Newton Abbot Devon

Contents

	LIST OF ILLUSTRATIONS	7
	INTRODUCTION	9
1	The Volunteer Force	11
2	Books to collect	26
3	Uniforms	34
4	Headdress, badges, buttons and equipment	50
5	Medals and tokens	60
6	Weapons	66
7	Rifle Volunteer Corps and their successors	76

APPENDICES:
1	Officers' rank designations for rifle regiments current in 1860	135
2	Volunteer Force Regulations—19 January 1861	136
3	Officers' rank designations—Volunteers 1881	137
4	Some books for the collector	138

BIBLIOGRAPHY	141
INDEX	143

List of Illustrations

PLATES

Imperial Yeomanry Volunteer	17
Surrey Yeomanry shako (*Wallis & Wallis*)	18
Bloomsbury Rifles helmet	18
Worcestershire Yeomanry helmet (*Wallis & Wallis*)	35
London Rifle Brigade helmet	35
Cinque Ports coatee (*Wallis & Wallis*)	36
Inspector of Volunteers and Yeomanry coatee (*Wallis & Wallis*)	36
Jacket of 2nd Hampshire RV	53
Jacket of Lothian and Border Horse (*Wallis & Wallis*)	54
Volunteer bayonets	71
Officers' pouches and belts (*Wallis & Wallis*)	71
Medal and Volunteer buttons	72
Helmet and shako plates	89
Cap badges	90
Pouch and glengarry badges	107
Cap badges	108

Unless otherwise acknowledged, illustrations are from the author's collection

LINE DRAWINGS IN THE TEXT

Title page of the 'little green book'	22
CIV lapel badge	24
Recruiting leaflet	28
Militia uniform—OR	37
Militia uniform—Officer	38
Cuff lace patterns	40
A Volunteer cartoon from *Punch*	42
Rifle Volunteer uniform 1859	43
Rifle Volunteer uniform 1863	44

LIST OF ILLUSTRATIONS

Shoulder title 5 Bn Royal Sussex Regiment	47
Shoulder strap 1VB Hampshire Regiment	48
Cap badge of 13 Bn The London Regiment	55
Cap badge of the Royal Devon Yeomanry	55
Cap badge of the East Riding Yeomanry	56
Crests of the Yeomanry in 1897	57
Boer War Volunteer's Medal	62
Bayonet marking	67

Introduction

Collecting militaria in all its forms is now the vogue; from cloth formation signs, buttons and the more common cap badges at one end of the scale with values ranging from a few pence to a few pounds, to the rare items for the wealthier collectors such as early nineteenth-century headdresses and flintlock weapons at the other. Books have helped to stimulate the interest in militaria, and although the specialists in the Territorial Army units which were formed in 1908 can find information from recent publications, the collector will search in vain for a recent comprehensive work dealing with the items of equipment used by British Volunteers. With the exception of World War II material, that which has been preserved in the largest quantity relates to the British Volunteer movement before the turn of the century. This book sets out, against a brief outline history of the Volunteers, to help collectors to identify the wide range of articles offered for sale by dealers and seen frequently in the auction rooms.

The author is indebted to Peter Huntley, who prepared most of the illustrations and to Wallis & Wallis for so kindly granting permission to reproduce their photographs.

R. J. WYATT

1: *The Volunteer Force*

IT IS NOT KNOWN WHEN ENGLISHMEN WERE FIRST CALLED UPON to defend their country against invasion, but by the eighth century Alfred was demanding service in his Fyrd from every male subject aged from sixteen to sixty, against attacks from the marauding Danes and Vikings. These levies supplemented the small number of trained professional soldiers when invasion was imminent, and were drawn from the freeholders and the ceorls who worked on their land. A different commitment also existed in the form of the feudal levy, but this duty was owed to a feudal lord or king by a tenant in return for the tenure of the land. In this way it was possible to raise an army for a private war, and even for service abroad, but it was not primarily for the defence of the country.

After the Norman conquest, William the Conqueror and his son William Rufus both resorted to the Fyrd to help repel the raiders from Scotland and Wales. On many occasions this form of the militia was misused, much to the annoyance of those who were embodied, and men were sent abroad for long periods. When, a hundred years after the Conquest, a French invasion was feared, 60,000 men were available as a national levy. In 1181 Henry II's first Assize of Arms caused every man to be classified by a jury and, according to his estate or property, to provide himself with the appropriate arms. He then had to swear on oath to be faithful to the king and to use his weapons only in his service. Twice a year constables inspected the equipment at what were called 'views of arms and armour'. Under Edward I's Statute of Winchester of 1285 'all men of every rank and degree, between fifteen and sixty, shall keep in their houses harness for to keep the peace'. They were all assessed as under previous statutes according to the value of their lands, goods and incomes, the wealthiest providing horses and armour and the poorer classes spears, bills and pikes, and as many as possible with longbows and arrows. In the hands of the British the bow proved an impressive weapon which all men were encouraged to handle with expertise. Edward III's army which set out to conquer France was raised under a new principle because the feudal system had proved unsatisfactory in obtaining recruits for service outside Britain. In the new system, the owners of property had to provide and equip complete units at their own expense, dependent upon the rate at which their worth was assessed. Later, Commissions of Muster appointed in the counties collected men together into bands for training and from these forces were developed the train bands.

When Henry Tudor defeated Richard III at Bosworth and became Henry VII in 1485 he had to re-establish the military system which had been modified to suit the old nobles during the previous wars. Firearms, which were first used in Europe in about 1325, were becoming increasingly more important than the bow, so the levy was in need of a major review. Henry's policy was to remove the militia from the control of the barons so that they were loyal only to himself, and each county had to provide its quota when called upon in times of emergency. He was careful also to monopolise all the available artillery, and to develop the arquebus. This new weapon made heavy body armour superfluous and tactics which demanded close combat and hand-to-hand fighting were altered during the reign of his son, Henry VIII. During the reign of Mary, the militia was partially re-armed under the lords lieutenant who appointed deputies and officers responsible for national defence in each parish. Enlistment was primarily by volunteering, but as this never sufficed to meet the demand the balance had to be made up by forcible impressment. Units, or bands, consisted of from 150 to 200 men under two captains and an ensign, and they were mustered from time to time for training and in order to suppress riots. Philip of Spain's attempted invasion in 1588, the Spanish Armada, led to the calling out of the entire armed force of the nation, altogether more than 120,000 men. It is perhaps just as well that, thanks to the navy, their services were not required, because they were ill-equipped and poorly trained soldiers.

The period of confusion and divided loyalties which existed at the time of the Civil War was caused partly by Charles I's wish to control the militia and to appoint its officers. Both he and Cromwell impressed men in the parts of the country which they dominated. The loyalist army consisted mainly of mounted loyal gentlemen volunteers, and the parliamentarians recruited the train bands in the cities, particularly the London bands who were such staunch supporters of parliament. Men of the militia joined General Monk when he marched to London to enable Charles II to return to England. At Blackheath, the new king met Cromwell's highly effective army, accepted its loyalty, and marched to the city to be cheered by thousands of militiamen. Charles' small private army, unsupported by public funds, consisted of only five regiments and there was still a need for the militia whenever there was threat of an invasion. The London bands and the county militia were rushed to the defence of London when the Dutch fleet sailed up the Thames, and the king assumed full control of the second-line forces.

William of Orange finally persuaded parliament to legalise and finance the regular or standing army, but the militia was still required to swell the ranks because the permanent force was far too small to be of much use on its own. Nearly 100,000 men were available for muster in 1715 and, after the invasion panic which ended at Culloden, the Militia Act of 1757 reorganised the reserve force yet again, men from 18 to 45 being chosen by lot to serve for three years unless they could raise £10 for a substitute. It was restricted to Protestants, and Catholics were not eligible for the militia until 1802. Provision was made for volunteers to be accepted in place of chosen men, and for captains of companies to augment their forces by enrolling volunteers if ever the militia were to be ordered out for actual service. By 1778, volunteers could form an integral part of the militia at any time, either by being included in existing units or as separate volunteer companies. Some of the militia regiments had as many as fourteen volunteer companies.

The first truly separate force of volunteers was raised when

England was again weak and threatened in 1782, but the many corps then formed were disbanded at the end of the war of American independence. Many of the officers and men who had volunteered then, once again offered their services twelve years later when the threat of invasion was as imminent as it had been two hundred years before. This time it was the French, not the Spaniards, who were preparing to attack the British Isles in the wars of the French Revolution which followed the overthrow of the French monarchy in 1789. The war did not begin until 1793 and the government first raised volunteers to guard the country in 1794 by augmenting the militia once again. This scheme failed completely because neither the gentry nor the workers were at all keen to be associated with the militia. It was only the poorest and roughest who served in the compulsory militia, those who could not find the money for a substitute and those who took the money to stand in for those who could afford to pay it. Militia men were looked upon as rabble and it took a very real threat of invasion before any self-respecting volunteer would wish to join them on a voluntary basis. This was realised by the administration and a new act, the Volunteer Act of 1794, appeared. It allowed men to 'voluntarily enrol themselves for the defence of their counties, towns or coasts, or for the general defence of the kingdom during the present war'.

An attempted invasion by thirty-eight French ships in Bantry bay, which failed because a gale prevented the troops from landing, led to the formation of a separate volunteer force in 1798 called the Armed Associations. Lords lieutenant were directed to prepare lists of all men from 15 to 60 who could be 'armed, arrayed, trained and exercised for the defence of the kingdom'. Very much along the lines of the Home Guard during the last war, they were designed to harass the enemy rather than to fight him in battle, they were not subject to military discipline and were not obliged to serve outside their own parishes. Including Ireland, the volunteer force of the period numbered no less than 410,000. Most of the volunteer corps and Armed Associations were disbanded after the peace of Amiens was signed in 1802, only to be revived in the following year when the war against Napoleon was renewed. Those who were not enrolled in any of the volunteer corps were to become special constables to aid the civil powers. The Levy *en masse* Act of 1803 provided for a form of conscription which divided the entire male population between 17 and 55 years into four classes for the defence of the country; however, there were so many volunteers that the Act became superfluous. Of the 500,000 liable to serve, 420,000 men volunteered; by the end of 1803 the number had risen to 463,000. All were subject to the Mutiny Act and all had to agree to march to any part of Great Britain in the event of an invasion or the appearance of an enemy in force upon the coast, and to assist in the suppression of any rebellion or insurrection arising at the time of an invasion. Lord Castlereagh's Local Militia Act of 1808 varied the temporary home defence force yet again, by producing by ballot over 200,000 men from the 18 to 30 years age group. No person drawn in the ballot was allowed to present a substitute and even a serving volunteer had to become a local militiaman for four years unless, if his income was less than £100 a year, he paid a fine of £5. Not surprisingly, virtually the whole of the volunteer force was transferred into local militia. In the first year, 250 regiments were raised and by 1812 there were 214,418 men serving, with another 68,463 true volunteers. The expected French invasion never came and in 1816, a year after Wellington's victory at Waterloo, the Local Militia Act was suspended and over the next two or three years all but a few of the volunteer corps were disbanded.

Britain entered into a long period of peace, even the Crimean War presented no threats of invasion, and it was fear of the French again which led to the great volunteer revival. Against the background of the defeat at Waterloo which was still within living memory, French animosity towards Britain was intense. At home there was no love for the French or for their Emperor Napoleon III with his large fleet and a standing army of half a million men. Feelings were intensified in 1858 by an Italian named Orsini. This man and his friends made bombs in England which they threw at Napoleon as he stepped from his carriage at the Opera in Paris. The emperor escaped injury, but eight people were killed and hundreds more injured when the crowd panicked. To counter the French attitude after this event, the British once again became hostile towards their old enemies and this, combined with feelings engendered by the poor state of the regular army, forced Lord Derby's government to accept the services of volunteers once again.

Apart from the Honourable Artillery Company, another corps which had served during the Napoleonic Wars and escaped subsequent extinction was the Duke of Cumberland's Sharp-shooters, the formation of which dated from 1803. Self-supporting and well organised, it had its own rifle range, armoury and drill ground, and changed its title after 1815 to the Royal Victoria Rifle Regiment. The regiment operated on the lines of a rifle club and started recruiting volunteers in 1853; by 1858 it had fifty-seven men, and this rose to 800 by the middle of the following year. A volunteer corps was formed in Exeter, the 1st Rifle Volunteer Corps, in January 1852, and its services were accepted by Queen Victoria in March, the first commission in the revived corps going to Capt Denis Moore, then town clerk of Exeter. In the same year the citizens of Liverpool banded together and in 1855 formed the Liverpool Drill Club.

After the attempt on the French emperor's life, General Peel, the Secretary for War, sent out a circular to all lords lieutenant outlining the government's thoughts on the formation of a defensive volunteer corps. It was proposed that officers should hold commissions granted by the lords lieutenant, that all members should take an oath of allegiance and should be liable to call-out in the event of invasion, the appearance of an enemy on the coast, or in case of rebellion, and that they should then receive regular pay and be subject to military law. Fourteen days' notice was required for a resignation and drill was to consist of eight daily parades every four months. All members were to be exempt from the militia ballot. The entire expense of the new volunteer force was to be met from their own funds,

Page 17 *A Volunteer of the Imperial Yeomanry dressed for a journey to South Africa in 1900*

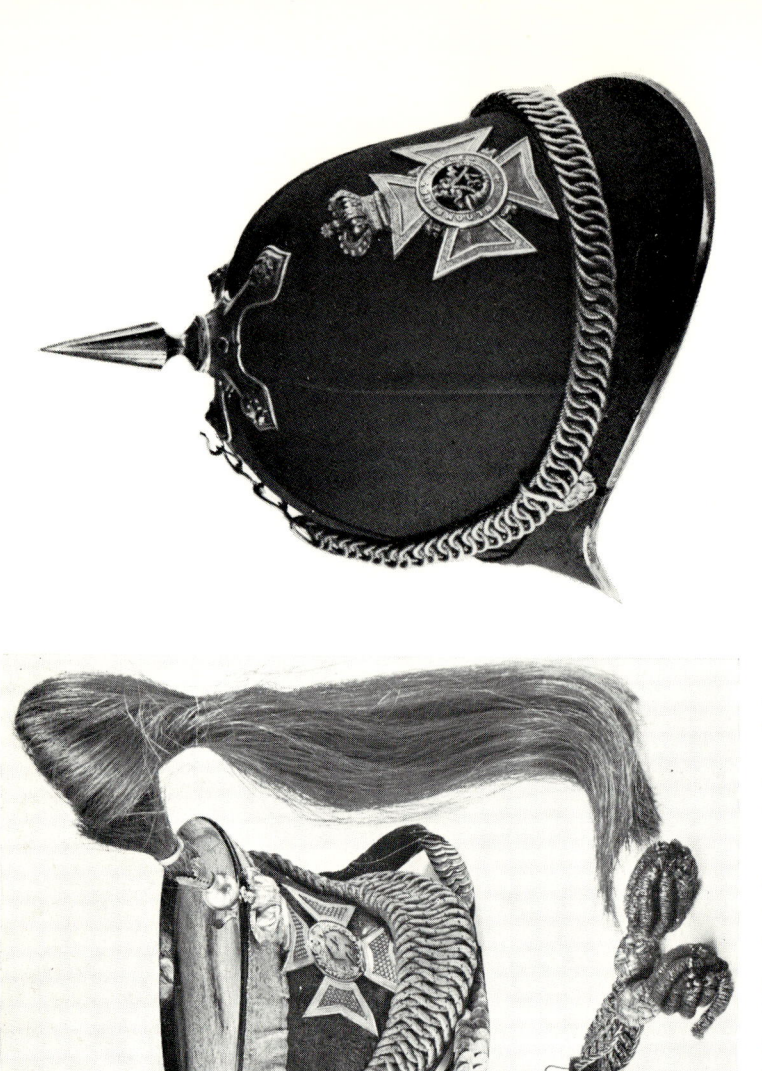

Page 18 (left) *Officer's bell-top shako of the Surrey Yeomanry c 1833;* (right) *home service pattern cloth helmet of an officer in the Bloomsbury Rifles c 1885*

including the provision of uniforms, arms and equipment. It is not clear how the government expected labourers and artisans, with scarcely sufficient funds to provide enough food for themselves, to be able to join a force at their own considerable expense. Some months later the authorities grudgingly agreed to issue government 0·577 muzzle-loading Enfield rifles to a quarter of the men in each volunteer corps.

It was intended that a corps should consist of a captain, a lieutenant, an ensign and about 100 men to act as a local force to harass the invading enemy's flanks. Sub-divisions were permitted to have as few as thirty men commanded by a lieutenant with the assistance of an ensign. Although provision was made for the grouping of companies into battalions, it was made clear that individual corps were not to lose their separate identities. They were all worried about their precedence, which was regarded as of next importance only to the style of dress which they adopted. The volunteer force in a county ranked in order of precedence in accordance with the date of the formation of the first individual corps in that county. Matters became more complicated because the artillery volunteers were senior to the rifle volunteers and had their own order of precedence. We find, therefore, that whilst the first rifle volunteer county was Devonshire, the first artillery volunteer county was Northumberland. In each county the separate corps ranked in order of the dates of their formation.

Finances to form and equip the rifle volunteer corps came mostly from public subscription. Non-effective members were enrolled and paid entry fees and subscriptions to defray the cost of the equipment supplied to those who undertook to join, but could not do so at their own expense. In Hampshire, for example, a number of the residents attended a public meeting called by the mayor of Winchester in the Guildhall on 26 May 1859. As it took place at noon on a weekday, it consisted entirely of local notables and tradesmen. Those present decided that it was desirable to form a local rifle corps in the city and its neighbourhood, and that a committee should be appointed to obtain names, addresses and occupations of fifty men wishing to join. As a rifle and uniform then cost between £6 and £7, an annual subscription of £1 was proposed and donations were sought from any person wishing to give support. Those who joined were asked to state if they would provide their own uniforms. Enrolment started on 30 May, by 1 June there were forty-three recruits of whom thirteen agreed to purchase their own equipment and fifteen could pay for uniforms but not rifles. At the first general meeting of the corps on 21 June all the officers and the sergeant-major were elected, a practice which was strictly prohibited afterwards. A total of £300 had been contributed to the funds. The Winchester volunteers were accepted by the authorities on 23 September with an authorised establishment of a captain, a lieutenant, an ensign and 100 other ranks. They ranked first in the county, Hampshire being given forty-fourth place in the county order of precedence.

Interest in the revived volunteer movement was much greater than had been expected. In the first two years men joined at the rate of 7,000 a month; in 1860 there were 119,000 and by May 1861 there were no less than 170,000. They included men from all walks of life, from peers to peasants, the former often subscribing to meet the cost of the uniforms of what was called the labouring class. The original idea of a force consisting of small companies and even smaller sub-divisions had to be revised because it was impossible to control and train so many independent corps. There were sufficient companies in almost all the counties to form battalions. Edinburgh, for example, had in its corps, the 1st (City of Edinburgh) Rifle Volunteers, more than seventy officers formed into what was called a consolidated battalion. In Hampshire, the 1st, 11th, 13th, 15th, 18th and 21st Corps, which were scattered around the county, were formed into the 1st Administrative battalion Hampshire Rifle Volunteers. In 1862, 48,700 rifle volunteers were organised into eighty-six consolidated (city and town) battalions and 75,500 into 134 administrative battalions. When grouped together in this way the volunteers could appoint an ex-regular officer as an adjutant, to assist in both administration and training. Initially they were paid by the volunteers, but from 1863 the army became responsible and insisted that they should retire at the age of sixty. In 1878 all adjutants were regular officers seconded to volunteer battalions for periods not to exceed five years.

Sir Charles Napier was convinced that it was undesirable to allow service in the volunteers to become irksome; he felt that all the drill necessary for them to know could be taught in six lessons, falling-in, marching, arms drill, firing by platoons, squad drill in two ranks and company drill, and this was all contained in a little green book *Drill and Rifle Instruction for the Corps of Rifle Volunteers*, first published in 1859. Full regulations for the volunteers were published in 1861 to consolidate the bewildering list of orders, memoranda and circulars which had been issued since 1859. Here a full complement of arms was allowed at government expense to each corps, together with free ammunition, and camps were set up at Aldershot and Shorncliffe for annual training at battalion level. The practice was extended in the following year when provision was made for camps to be formed in any part of the country.

Volunteers had now become such an important factor in national defence, with a strength of over 162,000, that in 1862 a Royal Commission was appointed to 'enquire into the character and composition of the force'. This recommended that for every man who attended nine drills a year, six of which were to be battalion parades, a grant of £1 should be paid for his subsidy, with a further grant of half that amount for each man who had fired sixty rounds and passed out as a third-class shot. Under the regulations of 1863, any volunteer corps or administrative regiment was permitted to form a cadet corps of youths from twelve years of age and upwards. The first cadet corps had been started in 1860 at the Royal Masonic School in Bushey and was attached to the London Rifle Brigade. Eton, Harrow and Winchester soon followed. A special contingent of what subsequently became the 1st Cadet Bn KRRC in 1894, served in the South African

DRILL AND RIFLE INSTRUCTION

FOR THE

CORPS OF RIFLE VOLUNTEERS,

By Authority

OF

THE SECRETARY OF STATE FOR WAR.

FOURTH EDITION.

LONDON:
PRINTED BY GEORGE E. EYRE AND WILLIAM SPOTTISWOODE,
PRINTERS TO THE QUEEN'S MOST EXCELLENT MAJESTY.
FOR HER MAJESTY'S STATIONERY OFFICE.

PUBLISHED BY
W. CLOWES AND SONS, 14, CHARING CROSS, S.W.

1859.

[Price 6d.]

Title page of the 'little green book', first published in 1859 war as part of the City Imperial Volunteers and was granted the only cadet force battle honour—South Africa 1900–2.

Growth in the size of the volunteer force more than kept pace with the increase in population and the figure of 153,500 was reached in 1868, but the number declined steadily until it had fallen to 130,600 in 1873, when again there was a revival of interest. A steady rise continued until 1886, when there were over 174,000 volunteers with a trailing off again for a few years afterwards as a result of the introduction of more stringent regulations. In 1867 another committee was formed to consider the adequacy of the capitation grant. It was reported that heavy personal expenditure fell upon commanding officers and that one captain stated that his company had cost him £500. The report, which recommended that the allowance be doubled, fell on deaf ears and the grant was not increased. Up to 1871, jurisdiction of the volunteers, including the granting of commissions, was the responsibility of the lord lieutenant of the county, but this was then changed to become the responsibility of the Crown under the control of the Secretary of State from 1872. The force became better trained and more closely associated with the regular army; every corps had to be attached to and form part of one of the seventy brigades or corps in the army, under the control of the officer commanding that formation.

The year 1881 saw a complete re-organisation of the British Army, in which all the old numbered regiments were re-formed into territorial regiments with two regular battalions, one at home and one serving abroad. Two militia battalions were attached to the county regiments, as were the volunteers, all the individual units being numbered consecutively from 1, the militia and the volunteers being kept separate. First administrative battalions in a county therefore became first volunteer battalions (1VB) of the re-formed county regiment. There were 215 volunteer battalions, the Queen's Rifle Volunteers, Royal Scots, having three battalions and the remainder being individual county battalions. Sizes varied from twenty-three companies for the 3VB Welsh Regiment to two companies on the Isle of Man and a solitary company at the Bank of England, the majority of the battalions consisting of from eight to twelve companies. This form of organisation for the volunteers continued until after the turn of the century.

The activities of the rifle volunteers during the nineteenth century were not at all spectacular. Two great reviews took place before Queen Victoria, one in 1860 in Hyde Park at which 18,450 men were on parade, the other at Windsor Park in 1868 with 26,953 troops. Then there were the annual rifle-shooting meetings at Wimbledon from 1860 until the ranges at Bisley were first used in 1890. With perhaps one exception, volunteers were not permitted to take part in military campaigns. In 1867, 1,600 members of the Post Office were sworn in as special constables at the time of the Fenian scare; the force was so well organised that it was decided to transform it into a rifle volunteer corps with an establishment of 1,000 and in 1868 it became the 49th Middlesex RVC, to be altered to number 24 during the re-organisation of the Middlesex corps in 1880. When it was decided to send an expeditionary force to Egypt in 1881, the authorities approved the recruitment of a special volunteer Post Office Army Postal Corps from members of the 24th Middlesex. Two officers and fifty men embarked for Egypt in August and on their arrival set up army and field postal units. At the battle of Kassassin on 9 September members of the 24th Middlesex gained the doubtful privilege of being the first British volunteers to come under enemy fire, and at the end of the campaign were awarded the Egypt Medal and Khedive's Star. The Field Telegraph Corps of the same volunteer unit went out to Egypt again in 1885 and served at Suakim.

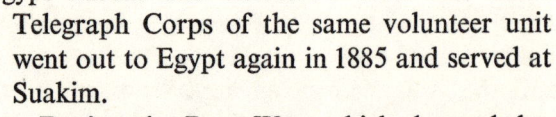

During the Boer War, which demanded a much larger force than was at first thought necessary to put down a few undisciplined Boer farmers, both yeomanry and volunteers offered their services. The former became Imperial Yeomanry, while the rifle volunteer

Lapel badge sold to raise money for the City Imperial Volunteers in 1900

detachments were formed into Volunteer Active Service Companies to be attached to their county regiments serving in South Africa. Both took part in most of the campaigns of that war.

The Territorial and Reserve Forces Act of 1907 changed completely the organisation of the second-line force. Twenty-three of the old militia regiments were disbanded and those which remained were transferred to the Special Reserve which was to be the first line of the army reserve. The term 'Yeomanry' was still applied to the volunteer cavalry regiments, but they were no longer to be a separate auxiliary force; together with the volunteers, they became the Territorial Force and later the Territorial Army. The entire cost of this new army was voted by parliament each year and it was administered by the county Territorial Force Associations. Volunteer battalion prefixes were removed from the titles and units became numbered battalions of their county regiments.

2: *Books to collect*

ON 1 JANUARY 1836 A HORSE GUARDS LETTER WAS CIRCULATED instructing all units of the army to keep official records in the form of regimental histories. Further details were contained in the *King's Regulations* of 1837, in which more precise directions were given as to the types of event to be recorded.

> An historical account is to be kept in every Corps of its services, etc; stating the period and circumstances of the original formation of the regiment; the stations at which it has been employed, and the period of its arrival and departure from such stations; the badges and devices which the regiment has been permitted to bear, and the causes on account of which such badges and devices or other marks of distinction were granted, are to be stated; also the dates of such permission being granted. Any particular alteration in the clothing, arms, accoutrements, colours, horse furniture, etc, are to be recorded and a reference made to the date of the orders under which such alterations were made. The various alterations which may be made in the establishment of the regiment, either by augmentation or reduction, are also to be stated in this book.

From this, one would expect there to have been preserved a wealth of valuable historical records giving details of the activities of all British forces. Unfortunately, this is not the case so far as the volunteers are concerned, and in all probability this is because the instructions were pre-dated by another regulation of 1822 which was brought out at a time when the infantry volunteers consisted of only two units, the Honourable Artillery Company and the Duke of Cumberland's Sharp-shooters, which later became the Royal Victoria Rifles. It was not until more than thirty years later that the Rifle Volunteer corps were formed, and the units and the War Office both seem to have felt that the regulations applied only to regiments already in existence and, in the view of the officials in London, hardly concerned the mass of part-time amateurs who were volunteering to serve without pay and to clothe themselves in quasi-military uniforms at their own expense. Consequently, information is not easy to find when studying the history of the volunteers.

Local newspapers of the mid-nineteenth century, however, are more helpful. They seem to have been less anxious about what would be regarded today as news than with printing in full everything which occurred in the district, however trivial. Volunteers parading in uniform were given columns, in some cases every man's name was recorded for posterity, and details of changes in uniform, weapons and equipment were provided. Copies can be consulted either in local reference libraries or at the publishers' offices, but if all that is required is confirmation of the date of the introduction of a particular badge or a style of uniform, the inquirer should be prepared to spend several days over the research, because local newspapers are seldom indexed.

Collecting items without being able to date them or place them in their correct historical period, can be frustrating. The hobby is enhanced by perspective. To obtain an old volunteer waist belt inscribed with the name of the original wearer may be satisfying; to know from the inscription on the clasp that it is associated with the county in which one lives, and from the shape of the crown that it is Victorian, may be sufficient for most collectors. Others will wish to know more. A little research could reveal the rank of the original owner, exact dates of his service and details of the battalion or independent corps in which he served. It is not necessary to be an historian to find this out, but it cannot be done without reference books. Books can be either complementary to other aspects of collecting, in which case they may be borrowed or referred to at museums or in libraries, or they can form a collection in themselves that it would be impossible to complete in a lifetime. Old books are not only paper with printing, or mere sources of information; part of their fascination stems from the fact that they have been owned by other people, people who are quite likely to have had a close association with the subject matter which they contain.

Rare military books, particularly early publications containing hand-coloured illustrations, are very expensive. Only the

London Rifle Brigade

Board of Patrons.

THE COURT OF ALDERMEN OF THE CITY OF LONDON.

President:—THE RIGHT HON. THE LORD MAYOR.

Lieutenant-Colonel Commandant:
LIEUTENANT-COLONEL LORD BINGHAM,
(*Late Rifle Brigade*).

Head-Quarters and Drill Hall: 130, Bunhill Row, E.C.

Telegraphic Address: "PARADING, LONDON."
Telephone No.: 9962 WALL

South Africa, 1900-1902.

Primus in Urbe.

Qualifications for Membership.—5-ft. 6-ins. in Height; 33-ins. round the Chest; Age, not under 17. Every Candidate must be proposed by Two Members of the Regiment, and must be a natural born or naturalised subject of His Majesty. Any Member is at liberty to resign on giving 14 days' notice through the Captain of his Company to the Adjutant, upon returning all Government and Regimental property in his charge, and upon adjusting any Regimental claim that may be brought against him.

Annual Subscription.—£1 5s. payable half-yearly in advance. No Entrance Fee.

Uniform.—Black, with Black Facings; cost with Great-coat, about £5, which, by arrangement with Secretary, on enrolment, can be paid by instalments.

Parades.—On alternate Saturdays (during the drill season), after 3.30 p.m. When these are held in the environs of London special trains are provided. Evening Drills (in plain clothes) 6.30 to 7.30 p.m. Annual Week's Camp.

Army Signalling, Maxim Gun, and Ambulance Sections are attached to the Brigade.

Rifle Ranges up to 1,000 yards at the City of London Ranges, Rainham, Essex.

School-of-Arms (Professional Instruction in Fencing, Boxing, Gymnastics, &c.). Badminton, Chess, Cricket, Football and Swimming Clubs. Masonic Lodge. Regimental Journal.

Bands.—Professional Band under a Military Bandmaster, also Drum and Bugle Band.

Cadet Corps.—For Boys between 12 and 17. Merchant Taylors' School, King's College School and Headquarters Companies.

All particulars and Enrolment Forms can be obtained from the Secretary at Head-Quarters, 130, Bunhill Row, E.C.

Recruiting leaflet for the London Rifle Brigade c 1905

wealthiest collector can ever expect to own an original copy of the most prized volunteer book of all, *The Loyal Volunteers of London and Environs*, published in 1799 and containing eighty-six coloured plates by Rowlandson. This work, with the two supplementary folding plates, *Sadler's Flying Artillery* and *Expedition or Military Fly*, could not be bought in good condition today for very much less than £700. Many books likely to be of interest are much less costly, and within the range from £3 to £10 will be found most of those available from even specialist booksellers. One of the joys of collecting books is that they are to be found in such profusion and in so many places; junk shops, jumble sales, attics, house sales and small auctions are only a few of the more obvious sources. A collection will never be complete, for even if all known gaps have been filled, it is still possible ten years later to find a book the existence of which had been previously unknown. It is advisable to buy a small indexed pocket book, and to study as many booksellers' catalogues as possible. From these and the bibliographies in military works of reference, a list should be compiled, alphabetically by author, in the indexed book, giving the full title, date, edition, details of illustrations, dealers' prices and dates. Carry the book with you and it will be found invaluable as a guide to titles and a reminder of their degree of scarcity and current value.

It is unlikely that, concerned as he is with a vast range of secondhand volumes, a bookseller will ever be able to gain as much knowledge as the collector whose concern is primarily with one subject. The dealer cannot hope to specialise in a branch in which only a few dozen books a year are handled, therefore he is more likely to buy well and be able to pass on the benefit to someone with a wider knowledge. Booksellers are also very helpful people; by their nature they encourage browsers. Very often they have time to spare, they attend sales and read all the catalogues and can advertise in their own trade journals. In common with most businessmen, they like to encourage regular customers. If you let them know your interests and call in from time to time, they can be guaranteed to find something; but it must be remembered that patience applies in the world of secondhand books. Get used to the idea of being told that the very book you want was sold last week, or that it should be in stock when next you call.

Periods during which the rate of publication of volunteer books was at its highest coincide with the level of interest in the subject. Hundreds of books, pamphlets and official and unofficial instruction manuals appeared between 1797 and 1805, during the widespread involvement in the protection of Britain from the threat of invasion by Napoleon. One interesting official volume from this era is *A Manual for Volunteer Corps of Infantry*, the first edition of which dates from 1803. Although published by the authority of the Adjutant General, like most books and pamphlets containing instructions on drill and exercises, copies were sold by T. Egerton at The Military Library in Charing Cross, a district of London which is still associated with books but, alas, not to the same extent as in the past.

At that time, books were sold unbound; people who bought them were wealthy and usually had them bound in leather to match the other volumes in a library. It was not until the 1840s that it became common practice for publishers to bind in cloth as we know it today. The unbound books were in what secondhand booksellers call 'boards'; card covers usually finished with a coarse blue paper with a binding of buff or white paper around the spine. A small label was stuck to the back of the spine on which was printed a shortened version of the title. The purchaser would have the boards removed when the book was bound for him, and the binder would cut the edges of the pages which the printer had left untrimmed. A book in its original state is called 'uncut in publisher's boards'. As books are made up from folded sheets larger than the page size, it was necessary to split the individual leaves before the book could be read, a process carried out normally when the pages were trimmed for binding, but often done carelessly with a cheap book not worthy of a leather cover. An untouched, therefore unread, book in boards is referred to as being 'unopened'. Surviving examples in this state are scarce and can be expensive. A tooled leather binding in good condition also increases the value, often regardless of the contents of the book, particularly now that old bindings are sold by the yard as decoration.

During the first half of the nineteenth century, following the disbandment of the large reserve force after the finish of the Napoleonic Wars with the defeat of the French Army at Waterloo in 1815, only a handful of titles were published relating to the volunteers. The number increased dramatically with the great revival in 1859, fell off slightly in the late 1860s and 70s, and reached its peak between about 1880 and 1895. One essential book, *The History of our Reserve Forces, with suggestions for their organisation as a real Army of Reserve*, published anonymously in 1870, was written by G. A. Raikes who served in the 3rd West York Light Infantry Militia. A prolific writer who was also responsible for the excellent history of that regiment, *The First Regiment of Militia* which appeared in 1876, he complains about the lack of information on the reserve forces in the official *Army Lists* of the period. Although names and ranks of all officers were listed under the corps in each county, all reference to uniforms and badges was omitted and the order of precedence was incorrect. In *Queen's Regulations*, only the badges and uniforms of regiments in the regular army were described, and even the comprehensive *Hart's Army List* could only find room for the militia at the end of the volume, after the index.

Raikes's book is invaluable because in it he set out for the first time in tabulated form the order of precedence of the bulk of the reserve units, giving distinctive titles, headquarters locations, colours of uniforms and facings and authorised establishments of both officers and other ranks. In spite of his valid criticisms, the small monthly volumes published for the War Office by HMSO entitled *The Army List: Officers of the Army, the Royal Marines, the Militia, Yeomanry and Volunteers*, are important to the collector. The original metal boxes which contained uniforms and helmets had engraved brass plates on the lids giving the name of the owner, his rank and unit. By referring to the *Army Lists*, it is possible to date the contents of such a box. Later *Army Lists* are not expensive and are quite easy to find, but those of the Napoleonic War period are rarer and most of them suffer from all the omissions which annoyed Raikes in 1870.

Histories of many militia, yeomanry and volunteer units appeared in the 1880s and 90s. The best of them were prepared after long and detailed research and provide accurate information on the service of the regiments against a background of general reserve army history. Rarely is there any data on dress, badges or equipment, although some contain illustrations of uniforms. Many are nothing more than dull lists of annual camp activities, with officers' promotions and range scores. Regular units produced and published their own private regimental journals; in the 1880s some of the volunteer battalions were given space for their own notes but some were more ambitious. For example, in 1890 a section of the Hampshire Volunteers began a monthly paper called *The 1st Vol. Batt. Hants Regiment Gazette*, in common with all similar journals it is extremely rare and the collector would be very fortunate to find a copy. There were some general volunteer magazines, notably *The Volunteer Service Gazette*, which began in 1859 and changed its name to *The Territorial Service Gazette* in 1908, and *The Volunteer: The non-coms' and Privates' own*, a weekly paper the first issue of which appeared on 12 November 1898. Scotland had its own *Volunteer News* in the 1870s and there were some annual publications, *The Volunteer Rifle Corps Almanac* from 1860 and the *Territorial Year Book* from 1909. All of these books are rare.

It is essential to have a copy of *A Bibliography of Regimental Histories of the British Army* by A. S. White, who was librarian at the War Office. It was published by The Society for Army Historical Research in 1965 and copies are still available. Mr White and Mr E. J. Martin also produced a *Bibliography of Volunteering* for the journal of the same society in 1945. Much of the pleasure to be had from collecting books comes with the discovery of new titles, and the enjoyment of finding hitherto unknown sources of information. Prints and illustrations in books provide particularly valuable sources of information for the uniform collector, and *An Index to British Military Costume Prints*, compiled by the Army Museums Ogilby Trust, gives details of over 15,000 plates.

A list of books for the collector will be found in an appendix but because Mr White's work can still be obtained, no attempt has been made to provide an exhaustive bibliography. Instead, it merely lists some of the more interesting and useful works. In all cases, copies have been offered for sale by dealers in secondhand military books over the last four or five years.

3: Uniforms

ONLY A RELATIVELY SMALL NUMBER OF THE MANY THOUSANDS OF collectors of military bygones specialise in service clothing. Uniforms have survived in spite of the ravages of dampness and moth, and the apparent lack of interest appears to result more from the problems encountered in display rather than anything else. Unless there is sufficient space for glass-fronted cupboards built into the recesses of a room, it is not an easy matter to show uniforms to advantage. Cloth is affected by dirt, sunlight, moisture and insect pests; buttons and braid become tarnished when exposed to the air, so uniforms are best preserved either in metal boxes or in polythene covers in wardrobes. A tailor's dummy or an old full-size clothier's model can be dressed in a different uniform from the collection each month, provided it is kept out of the direct glare of the sun, sprayed with a moth repellant and brushed before it is returned to its polythene cover for long-term storage.

Because uniforms have the inherent disadvantage of being cumbersome, resulting in a greater supply than there has been a demand, prices are relatively low in comparison, say, with pre-1901 cap badges. It would not be impossible to buy an early nineteenth-century volunteer shell jacket for the price of a couple of Victorian helmet plates. Army uniforms have been reproduced by theatrical costumiers for use in films, plays and on television, and as some of these are old and show signs of wear it is not always easy to distinguish them from the genuine articles. Fortunately, the entertainment profession does not need volunteer clothing, so it is unlikely that any has been produced; if it can be established that a particular item is of volunteer origin, then it is almost certainly genuine.

The dress of the volunteer differed from that of his regular army counterpart right up to the formation of the Territorial Force in 1908. It varied over the years, sometimes the uniforms were alike and at others they bore very little resemblance at all

Page 35 (left) *Helmet of the Queen's Own Worcestershire Yeomanry c 1850;* (right) *shako of the London Rifle Brigade, black cock's feather plume removed to show badge, post 1901*

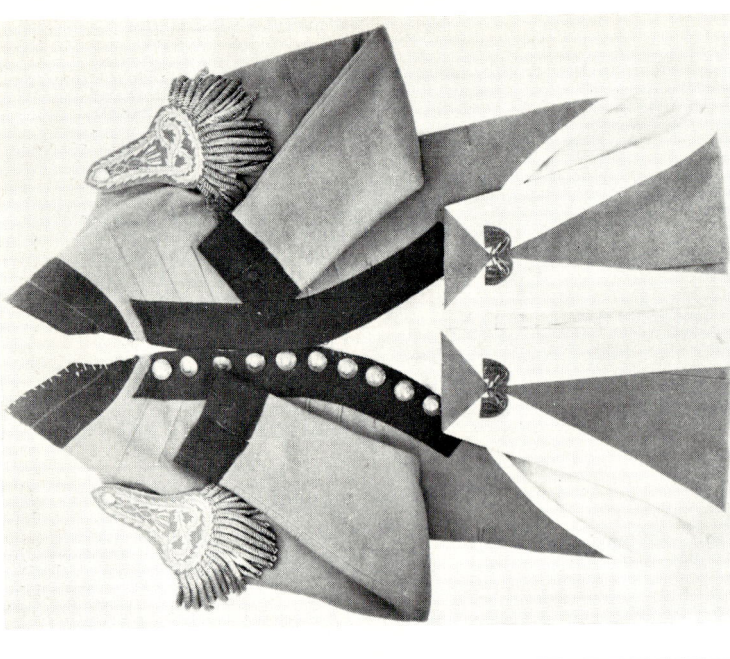

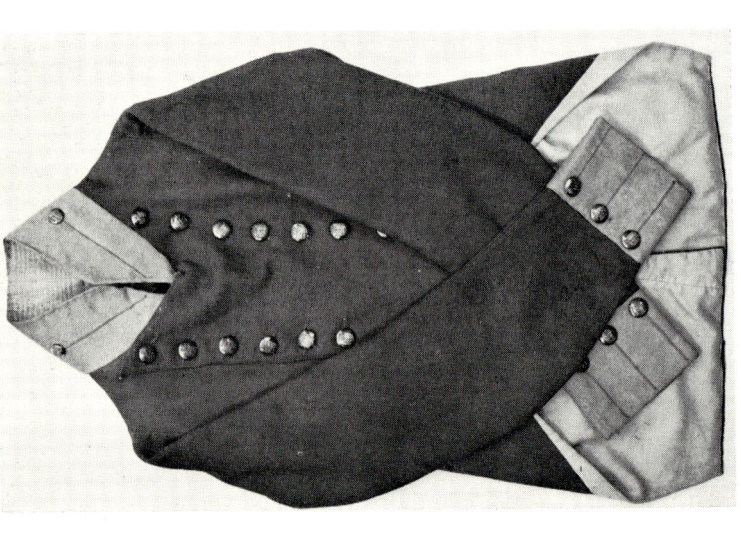

Page 36 (left) *Coatee of the Cinque Ports Volunteers early 1800s;* (right) *coatee of an Inspector of Volunteers and Yeomanry c 1795*

to the dress of the regular soldier; the descriptions of the various styles which follow will enable the collector to establish the authenticity and the date of most types commonly available. It must be remembered that volunteers were not always subject to the same stringent regulations which standardised the dress of the professionals. During certain periods either the units, or in some cases individual members, purchased their own dress; one of the reasons for joining was to be seen in a splendid uniform. As they were seldom worn, other than on parades and on Sundays and holidays, they lasted longer and were sometimes altered when styles changed, often combining features from earlier periods. Thus Edward VII crowned buttons may be found combined with Victorian collar dogs, Victorian tunics with original buttons, but Edward VII rank crowns on shoulder straps appear when an officer was promoted after 1901. Almost anything must be expected but, in general, a tunic can safely be dated from its earliest features.

1. *Rest your Firelock. 1st Motion*

Militia uniform—Officer Norfolk Militia 1759

On rare occasions, items of volunteer clothing of the period from 1790 to 1810 are sold by auction, without attracting very much interest. A helmet of the period will sell for three or four times the amount realised on a shell jacket, and although it would not be possible here to attempt to describe the very wide range of items in detail, the following brief descriptions of

C

some typical uniforms will aid identification. Styles of militia dress prior to the beginning of the war with France in 1793 for officers and other ranks are shown in the drawings on page 37 and alongside taken from the plates which illustrate *A Plan of Discipline, composed for the use of the Militia of the County of Norfolk*, published in 1759. Here the jacket was long and took the form of a light red overcoat with turn-over collar, covering a long buttoned waistcoat. Although the officer's jacket had about twelve buttons at the front, it was not designed to be fastened, but rather to hang loose like a cloak.

Long tails remained in fashion until after about 1803, and the buttoned waistcoat was gradually shortened as it became the custom to button the jacket. By 1794 the high collar was fastened at the neck, the jacket was buttoned about half-way down and curved away in shell shape to form the two long tails. Cuffs, collars and fronts were in almost any colour, but the jacket was most often red. The official volunteer regulations of March 1803 specified that: 'The whole to be clothed in red, with the exception of the corps of artillery, which may have blue clothing and rifle corps which

Position of an Officer carrying his Fusee on his right Arm.

Militia uniform—Other ranks Norfolk Militia 1759

may have green with black belts.' In a few years, double-breasted jackets, fully buttoned, came down to the waist with much shorter buttoned tails covering the hips and rear. Trousers for all these periods were of white buckskin or cashmere, tight and buttoned at the knee.

There is very little to interest the uniform collector between 1815 and the resurgence of the volunteers in 1859. On 12 May 1859 a War Office circular casually made the following statement on volunteers' dress: 'The uniform and equipment of the corps might be settled by the members, subject to the approval of the Lord Lieutenant.' Authority had failed to realise that here was the one problem they were going to have with their new protectors. Men were glad to serve without pay, but they wanted to choose their own style of uniform, one that did not bear a close resemblance to that of the regular army. Reasons are not hard to find. Soldiers were recruited from the lower classes of society and few volunteers would have been pleased to be mistaken for regulars. Another fear, albeit unfounded, was that if both forces were dressed alike, it would be easy for a government to mobilise the volunteer force and assimilate it into the army by Act of Parliament. Struggles between commanding officers and lords lieutenant were long and bitter, sometimes resulting in the resignation of a unit's leader.

At an early meeting in Winchester in 1859 of the 1st Hampshire Rifle Volunteers, a committee was formed to look into the question of a uniform for the new corps. One elderly gentleman brought the jacket which he had worn fifty years before as a young member of the Loyal Volunteers. It was inspected with much interest, but the specimen the members chose to adopt consisted of a dark green tunic, tartan trousers and a light cap trimmed with black braid. The lord lieutenant was not impressed. He had received another communication from the War Office advising him that the uniform adopted should be as simple as possible and 'that the different companies serving in the same county should be assimilated, and, although this point is left to the discretion of the volunteers subject to the approval of the lord lieutenant, it is considered that a recommendation on the subject would be an advantage'. Winchester's proposed uniform was rejected; whereupon their secretary wrote to the War Office claiming that the lack of agreement was detrimental to recruiting because many people would not join until they knew the dress of the corps. A volunteer had to pay for his equipment, so why should he pay for a uniform that was not to his liking? He also asked what powers a lord lieutenant had to reject a pattern chosen by a unit. In their reply, the War Office told him that a lord lieutenant was not empowered to enforce upon any volunteer corps a uniform to which they objected, or to withdraw his approval once given. But, on the other hand, volunteers were not to adopt any uniform 'of which the lord lieutenant may consider it necessary to express his formal disapproval'.

Agreement, by compromise on both sides, was reached in 1860, and the result was typical of most other early styles. The jacket was known as a patrol jacket, a single-breasted type without buttons, fastened with hidden hooks and eyes, and without pocket flaps, in light grey material with light green facings—a term used to denote the colours, special to the corps, of the cuffs and collars. Collars were square-fronted and fastened to the neck. These jackets were longer than used in the later tunics, being 27in long from the bottom of the collar to the skirt for a man of 5ft 9in. Cuffs were pointed, with $5\frac{1}{4}$in of facing material, with a 1in black braid edging. There were twisted black cords on each shoulder. On the patrol jacket, rank was denoted by sleeve and collar

Officers' cuff lace patterns 1880–1902. Lieutenant and Lieutenant-Colonel

braiding, stars and crowns. Captains wore an inverted vee 1in braiding on each sleeve, 8in from the cuff to the point, with a row of eyes outside, turned with a loop at the top, and with thin braid, called tracing braid, below the 1in braid with a loop turned from the point towards the edge of the cuff. A row of eyes was set below the braid at the top of the collar, and on each side there was a crown and a star of black silk embroidered on green velvet. Lieutenants and ensigns had a similar sleeve pattern using plain tracing braid without the row of eyes above the 1in braid, lieutenants with a crown only on each side of the collar and ensigns with a star.

Rank badges for non-commissioned officers consisted of a black silk crown on each sleeve above the cuff for a sergeant-major; a star in the same position for a QM sergeant; a device of crossed swords, crown and bugle enclosed in a wreath on each arm above one embroidered black silk and silver $\frac{3}{4}$in braid stripe for a colour-sergeant; three stripes of $\frac{3}{4}$in black braid on green facing cloth above the elbow of each arm for a sergeant; and two similar stripes for a corporal. Stripes were worn on each arm until 1881. Trousers were light grey with a light green $\frac{1}{4}$in stripe down the outside seams. In 1859 some rifle volunteer corps issued plain grey or black trousers, bought from subscribed funds, to members unable to afford their own equipment. Not surprisingly, these were worn for work and pleasure and scarlet braid was stitched to the outer seams, but this was easily removed so the practice of sewing scarlet piping into the seam was adopted. At about this period, or shortly afterwards, a number of corps followed the example of the Victoria Rifles and based their dress on that of rifle regiments of the regular army; details of officers' rank insignia are given in Appendix 1.

Unfortunately, the volunteer officers' childish, bordering on feminine, concern for their appearance in 1859 and 1860 resulted in ridicule from the press. *Punch* led the way by publishing cartoons and articles in almost every issue. Silk sashes were being worn over the uniforms, gold lace adorned shoulders and tunic edges, distinctions which were normally reserved for

UNIFORMS

VOLUNTEER (HE OF THE 'TASTEY' UNIFORM). "*And it's so comfortable and easy, that I shall most decidedly 'shoot' in it next Season.*"

Punch *cartoon of February 1860*

regular officers. Unauthorised swords were also being carried by all ranks and the War Office was writing continually to lords lieutenant requesting that they should 'signify the disapproval of the War Office' to commanding officers of the offending corps. Swords, it was pointed out, could only be carried by commissioned officers and sergeants when on duty or travelling to and from duty, and all lace had to be silver and of the standard army pattern. By the end of 1860 most of the initial problems had been solved, for, once the first corps in a county had been given approval the lord lieutenant was not empowered to revoke his consent, and simply referred any subsequent corps to the first so that styles could be assimilated. Button design and facing colours were left to the discretion of each corps. Full War Office regulations on the conduct of the volunteer force were published early in 1861, and the section dealing with dress is given in Appendix 2.

As one of the main objects of the volunteer movement was to train men in the use of the rifle, and as the members were so keen on displaying marks of distinction, proficiency badges were soon authorised. In 1861, an embroidered rifle was worn horizontally on the sleeve by the best marksman in a unit at ranges up to 300yd. A star was worn above the rifle by the best shot from between 350 and 600yd, and every

Rifle volunteer officer's uniform of 1859

man who was a first-class shot at 900yd wore two stars above the rifle. The individual first-class marksman with the highest score at 650 to 900yd wore yet another star. Each qualified musketry instructor was allowed to wear crossed rifles and a crown above.

Colours of the material used for uniforms were restricted in 1863 and, unless the lord lieutenant had already granted authority for a special colour, all arms in each county were to conform to one, with all units in an administrative regiment to be clothed alike. Sealed colour patterns were available at the Royal Army Clothing Depot at Grosvenor Road, Pimlico. Units were allowed two years in which to comply with the new standards; so by 1865 all but a few corps had taken the offer of the government and purchased cloth or tweed at cost from Pimlico in scarlet, white, blue, green or grey for tunics; green or grey for trousers and red, blue or green serge for frock coats with rifle green tartan for the trousers. Artil-

Rifle volunteer officer's uniform of 1863

lery blue or grey coat material was also available. Tunic design was more practical, the skirts being cut shorter and all braid now conformed to the regular army patterns.

Marks of efficiency—meaning in this context the completion of twelve months as a member of the corps, having carried out the prescribed number of drills and attended the annual camp—were introduced in 1864. The badge was an inverted chevron of ¾in silver lace worn on the right cuff of the tunic; later altered to a silver lace ring above the right cuff passing over all other sleeve ornaments. In 1872, one silver star was added above the ring for five years service and two stars after ten years.

Major changes in the colours and patterns of uniforms were made again between 1874 and 1877. Grey patrol jackets with coloured facings and black braid trimmings, five toggle fasteners similar to the type used on modern duffel coats, with black braid and loops across the whole width of the front, were considered to 'lack smartness' in the opinion of the 1st Hampshire Volunteer Battalion. In Portsmouth, the 2nd Battalion had discarded the old form for a buttoned, single-breasted tunic, but retained the grey colour. The 1st applied to change their dress to conform to that of the 2nd, but decided afterwards that it was not exactly what they wanted, cancelled the application and asked to be clothed in red with yellow facings, white metal buttons and silver lace. Current regulations were that, when adopting scarlet, volunteers would use the same facings as the militia of their county and that if a county had more than one militia regiment the Secretary of State was to decide which one should be used by volunteers. Hampshire Militia facings were black, and the uniform of the 1st Administrative Bn Hampshire RV in 1877, again typical of most others of the period, was as follows: scarlet tunic with black facings, with white piping down the front, at the bottom of the collar and at the top of the sleeve facings. Buttons were of white metal. Trousers were dark blue serge, infantry pattern, with scarlet welts sewn into the outside seams. The regulations of 1878 ordered that the various corps making up an administrative battalion were to be clothed alike by 1 April 1879, but that any groupings organised subsequent to 1874 should be allowed up to five years to conform.

With the widespread introduction of scarlet tunics in 1877, it was necessary to find a way to distinguish volunteers from regulars, so all volunteers were required to wear an Austrian knot of flat tape on each sleeve. Green uniforms had light green knots; blue uniforms, scarlet knots; scarlet uniforms, knots of facing colour, unless the facings were red, in which case knots could be dark blue or black. Units clothed in green used green facings of the same shade as the jacket, or of the colour of the knot.

At the same time, a further distinguishing mark was introduced, designations on shoulder straps. Each strap carried the initial letter or letters of the county (Hampshire being HS) and the number of the corps in the sequence in the county as given in the *Army List*. Particular care has to be taken when dating a uniform by the number on the strap and relating this to a town or district. In 1870, the 15 Hampshire Corps was centred at Yateley. Some years later, the captain, a Yateley man, retired and a new captain took over who lived a few miles away at Hartley Wintney. He was given permission to change the headquarters of the corps from Yateley and the unit became 15 (Hartley Wintney) Hampshire RV.

In addition to the county letter and its number, a corps also had to include the number of any administrative battalion to which it was attached. If it belonged to an administrative battalion of a different county, the letter of that county had to be shown as well as that of the county in which the headquarters of the corps was located. Luckily, this seldom applied or there might have been a need either to enlarge the shoulder straps or to decrease the size of the letters and numbers. Designations were in the same colour as the Austrian knot for green, blue or scarlet uniforms. For grey tunics, the same colour as the facings, or as the braiding or piping if the facings were grey. The shoulder straps were edged in tape of the same colour as the letters and numbers.

UNIFORMS

Efficiency insignia at this period consisted of a ½in wide ring of either silver lace, cloth or braid, on the sleeve of the right arm and passing under any other lace or embroidery. For five years' service, and each further five, a silver, silk or worsted star was added. Gold lace, gilt and brass ornaments were prohibited for other ranks. Sergeants in scarlet or blue uniforms wore three silver chevrons edged in red on the right arm; in grey uniforms, silver chevrons or in any other material authorised; green uniforms, light green chevrons. Proficiency badges, in the same material as the stripe, in the shape of a star, were worn above the chevrons. Sergeant instructors wore three chevrons on each arm with crowns above. Crossed muskets could only be used by NCOs with certificates from the School of Musketry, Hythe.

Officers' scarlet or blue tunics used silver cord or braid, edged with scarlet on the blue uniform. When in grey, the same silver cord or braid was used, but other materials could be authorised. With green uniforms, black cord or braid was applied, with a light green edging. When in grey or green, the rank patterns on the sleeves were to be the same as for regular rifle regiments. All ornaments were in silver with one important exception which was current until after 1900; gold star and crown rank badges were used where regular officers used silver.

In 1881 the 109 numbered infantry regiments were re-organised on a territorial basis. New titles were adopted, many pairs of regiments were amalgamated, numbers were dropped and names used, most of which associated the regiments with a county. Rifle volunteer corps became the volunteer battalions of the regular county regiments and it is not surprising that there were more changes in uniform to

White metal shoulder title of 5 Bn Royal Sussex Regt post 1908

aid the assimilation of the corps with their parent line regiment. With the exception of royal and rifle regiments, all English and Welsh territorial regular units adopted white facings and rose-patterned silver lace, the Scots yellow with thistle-pattern lace, and the Irish green with shamrock silver lace.

Generally, men in volunteer battalions wore the same uniforms as the regulars, with modified distinctions and with some restrictions with regard to collar badges. Other ranks dressed in green were to wear light green Austrian knots; in blue, scarlet and in scarlet, black. Shoulder straps lost their edge braiding and designations consisted of the number, the letter 'V' and the regimental abbreviation. For officers, the square section sleeve braid was replaced by flat braid to denote rank. Other details of officers' uniform are given in Appendix 3.

An alteration to the wearing of badges by marksmen was made in 1887. The best shot in a battalion wore crossed rifles and a crown on the left sleeve; in silver embroidery or cord if the uniform was green. This same badge was also used, but on the right sleeve, by the sergeant of the best company in each battalion. Each company best shot wore only the crossed rifles on the left sleeve, and all marksmen cloth crossed rifles.

Red tunics were still common after the death of Queen Victoria, with one or two minor alterations. Cuffs were squared instead of pointed, Austrian knots were removed, and the skirt flaps which had previously used two buttons and two lines of white piping, were altered in 1902 to a design with six buttons and three pointed edges. When

Shoulder strap of the 1st VB Hampshire Regt

the Territorial Force was formed in 1908, white metal shoulder titles were used, with the letter 'T' and the county abbreviation in curved form beneath. Khaki was adopted by some volunteer battalions in 1903 and the only way of distinguishing a volunteer from a regular by his uniform was by shoulder titles or, in some cases, cap badges.

4: Headdress, badges, buttons and equipment

PERHAPS THE MOST DESIRABLE ITEMS TO THE COLLECTOR ARE THE splendid helmets and shakos worn by the yeomanry during the nineteenth century. Although they are not difficult to find, they do command very high prices at auction and even higher values are placed upon them by some of the London dealers. Yeomanry volunteers, if they were not always wealthy themselves, were often commanded by rich landowners and as the regiments bought their own clothing they completed their colourful uniforms with elaborate and highly decorative helmets. They did not always follow the current regulation cavalry pattern and each regiment needs to be studied independently if a helmet is to be dated accurately. W. Y. Carman's *Headdress of the British Army—Yeomanry*, published privately in 1970, is the best book on the subject. Some early yeomanry helmets are illustrated in the plates, all of which have been sold at Wallis & Wallis auctions in recent years. The four most common basic types are the leather helmet surmounted by an ornate pressed brass crest, sometimes covered with a feather or an animal skin plume; the tall shako with either parallel sides or of the wide or bell-topped variety; the standard pattern leather or white metal regular cavalry helmet; and the fur busby adopted by many yeomanry regiments at the end of the last century.

Early infantry volunteers chose a less spectacular headdress. Until about 1796, they followed the militia dress styles, using a black three-cornered or tricorne felt hat of a type which had been popular for some fifty years. This was followed by the black felt cocked bicorne hat, at that period worn across the head in the French style, later to be worn fore and aft. Both these types are very scarce and all those which have survived now appear to be in museums. Later on in the Napoleonic Wars, infantry volunteers adopted a tall stove-pipe shako with thick wooden protection built into the crown.

HEADDRESS, BADGES, BUTTONS

From 1859, the date of the great volunteer revival, headdress tended to follow the infantry styles then current. Members of the 1st Hampshire RV wore forage caps, a form of pill-box hat with or without a peak and known in contemporary slang as a 'pork pie' or 'muffin'. It was in light grey cloth, 2½in high at the front and 4½in at the back, with light green piping to the edge of the crown, a green button on the centre of the top, with black tracing braid on each edge of a green band around the crown. The Hampshire Volunteers' hats had straight peaks. Some units adopted black or green cock's feather plumes which fell forward over the front of the cap. Forage cap is a confusing description, because it was the name given to the side hat used during the 1939–45 war which, when introduced in the 1890s, was known as the field service cap. The correctly named forage cap ended its days as the small pill-box perched at a jaunty angle and kept in place by a leather chinstrap, which went out of fashion in the early 1900s.

By 1878, regulations dictated the style of the forage cap for volunteer officers. When dressed in scarlet, a blue cap was worn with a band of black oakleaf braid, edged with the facing colour of the corps, the crown button was also in that colour. In green, the cap was green with a black braid band edged with green and having a light green button. Corps who chose grey uniforms had grey caps with a silver, black, grey or facing colour band and button. All these caps were without peaks. Hampshire volunteers first wore shakos in 1864, of a type similar to that worn by members of the London Rifle Brigade well into the present century, except that the earlier types employed a woollen pom-pom decoration in a leaf-shaped cup on a small metal ball, with silver braid on the crown and a white metal or silver-plated chin chain. In the 1870s, most corps replaced the shako with the soft woollen glengarry; Berkshire volunteers adopted it in 1873 and by 1877 the 1st Administrative Battalion in Hampshire were using a blue glengarry with a red tuft for both full and undress uniform. A hussar-type fur busby was also quite common, the 17th Middlesex RV first used one in 1871 with a tall standing feather plume. There were almost as many plume styles and colours as there were corps, but in 1891 units still wearing them had to conform to light green feathers or horsehair for the lower part of the plumes, when wearing green uniforms, or of the colour of the facings when scarlet or grey uniforms were worn.

In 1878, the British Army decided to adopt the home service pattern helmet, which was similar in general design to the Prussian pickelhaube, and is still familiar as the helmet now worn by most British police forces. It had a cork body covered with cloth in four seams, a pointed or rounded peak at the front edged with metal for officers and leather for other ranks, and a rear peak edged with leather. Officer's headdress had a metal strip down the back running from the spike to the rim of the rear peak. On top of the helmet a metal spike screwed into a cross-shaped base. A ball replaced the spike for helmets used by members of the Royal Artillery, Army Service Corps, Army Medical Staff and the Army Veterinary Department. At each side, a rosette held a metal chain made up from $\frac{3}{8}$in wide links on a leather backing strap lined with velvet. Volunteer helmets were covered in cloth of a number of different colours, all with silver-plated fittings for officers and white metal for other ranks or with metalwork having a dulled silver or blackened bronze finish. Cloth was either dark blue, green, rifle green or grey. Units were quick to adopt the new helmet, the 29th Middlesex ordering theirs from Firmin's as early as July 1879.

Volunteer helmet plates and headdress badges do not follow any strict pattern prior to 1881; early styles resulted from the whims and fancies of commanding officers, sometimes even being based on Napoleonic War volunteer designs, where a corps raised in 1859–60 felt that it had a link with one of the old disbanded units. The first forage caps and busbies mostly used the brass or bronze volunteer rifle bugle, quite often containing in the centre circle either the number of the county precedence or that of the corps within its county with a crown above on a separate corded boss. Shakos used either a crowned wreath containing the number within a garter scroll around which was

Page 53 *Sergeant's red jacket of the 2nd Hampshire Rifle Volunteers c 1880*

Page 54 *Colour Sergeant's dress jacket of the Lothian and Border Horse Imperial Yeomanry c 1902*

HEADDRESS, BADGES, BUTTONS

inscribed the county title, for example—Middlesex Rifle Volunteers; or a similar badge fixed on to a universal pattern star plate with eight points and a crown above where the top point should be. This same star plate was used on the home service pattern helmet, with a centre plate as mentioned above, or one in the shape of a Maltese cross with a ball on each point

Cap badge 13 Bn (Kensington) The London Regt

Cap badge Royal Devon Yeomanry Artillery

and a centre circle containing a stringed bugle. A larger black or bronzed Maltese cross plate without the universal star plate backing was also quite common throughout the nineteenth century.

When the numbered regiments of the regular army were given their territorial designations in 1881 and the rifle volunteer corps and administrative battalions joined them as county volunteer battalions, there was much greater standardisation in helmet plate design. A white metal or silver-plated star plate and crown was used, on to which was attached a circular helmet plate centre containing the regimental emblem and around which was inscribed the regimental title and, for example, '1st Volr Battn.' There were more ornate designs, such as the 2 VB Kings Own Scottish Borderers' plate shown in the illustrations. The helmet plate centres, with a small crown fixed to the top, were used on the glengarry. Badges can be dated by the shape of

D

the crown; the St Edwards crown was used until 1901 and is known as Queen Victoria's crown (q.v.c.) by collectors, having wider sides than the more compact Imperial or King's crown (k.c.) in use from 1902 to 1953. Up to the formation of the Territorial Army in 1908 many units used the small cap badges,

Cap badge East Riding Yeomanry 1920–56

mostly in white metal. They were similar in design to those worn by the regular units to which they were attached, but sometimes lacked the battle honours to which they were not entitled, the plinths containing them being left blank. Badges since 1908 are dealt with splendidly in such books as John Gaylor's *Military Badge Collecting*.

This hobby, one of the most interesting and rewarding pastimes, has now become expensive and in one way exasperating. The demand surpassed the supply, so enterprising dealers bought the original dies from some of the old badge makers and flooded the market with reproductions—they can hardly be described as forgeries—which are retailed at from £1 to £2 each. Only one or two volunteer items have turned up so far, but collectors should be very wary of any cap badges of the yeomanry regiments. They have all been reproduced in their thousands.

Most volunteers, particularly those dressed in grey and rifle green uniforms, wore brown or black patent leather shoulder cross-belts with a small pouch on the back. The pouch often had a bugle and crown and on the front of the belt either a glengarry badge or a special cross-belt plate, either circular or in the shape of a Maltese cross, between a lion mask boss from which hung chains connecting to a small whistle which clipped into a fitting

1897—1 Buckinghamshire; 2 Gloucestershire; 3 Derbyshire; 4 Lancashire Hussars; 5 Kent; 6 Somerset, North; 7 Kent, West; 8 Yorkshire Hussars; 9 Herts; 10 Devon, Royal North; 11 Oxfordshire; 12 Denbighshire; 13 Lanarkshire; 14 Cheshire; 15 Lanarkshire (Glasgow); 16 Pembroke; 17 Hampshire; 18 Notts, South; 19 Somerset, West; 20 Berkshire; 21 Devon, Royal 1st; 22 Montgomeryshire; 23 Westmorland & Cumberland; 24 Staffordshire; 25 Warwickshire; 26 Northumberland; 27 Yorkshire Dragoons; 28 Shropshire; 29 Middlesex; 30 Dorset; 31 Lothians & Berwickshire; 32 Notts (Sherwood Rangers); 33 Suffolk; 34 Leicestershire; 35 Lancaster's Own, Duke of; 36 Wiltshire; 37 Worcestershire; 38 Ayrshire lower down the belt. These were in white metal, silver plate or hall-marked silver.

The following accoutrements were authorised to be used by the Hampshire RV in the 1860s: Officers—pouch belt in black morocco leather, with silver-plated or silver lion's head whistle and chain, Maltese cross breast plate with Hampshire rose in centre, with a small silver or plated bugle on the pouch. Sword belt in black morocco leather, plated furniture, snake fastening to front, to be worn under the jacket. Sword of Rifle Brigade pattern, half basket hilt in a steel scabbard, with a black leather knot.

Staff Sergeants—pouch belt as for officers, but lion's head, whistle and chain to be bronze. Maltese cross with wreath and rose only in silver or silver plate, bronze bugle on pouch, the rose to be silver or plated. Sword belt and swords as for officers, but the scabbard to be black leather with steel mounts.

Sergeants—cross-belts to be plain leather with a 20-round pouch of patent leather at the rear, bronze lion's head whistle and chain, Maltese cross breast plate, bronze bugle on pouch, plain leather waist belt with patent leather frog and ball bag with bronze snake fastening to the front.

Corporals and below—as for sergeants, without ornaments to the cross-belt and pouch. Black leather leggings were authorised shortly afterwards.

By 1878, corps with blue or scarlet uniforms were to wear white waist belts and black pouches, with green uniforms black belts and pouches, and with grey dress black or brown belts and pouches.

Volunteer buttons present an interesting and as yet undiscovered field for the collector, the only difficulty being that they are very hard to find. The volunteers seem to have had a free hand with buttons, provided they were in black bone or white metal or silver they seem to have been able to choose their own designs. Many of the bone buttons used the bugle symbol or a circle with the county title above the corp's number, which was surmounted by a crown. Dating is possible from the titles adopted by the makers and marked on the flat backs of the buttons. The most popular manufacturers were Wright of Birmingham, who also made most of the regular army other-ranks buttons from 1855–71, and Firmin and Jennens. Volunteer buttons by Firmin are found marked Firmin & Son, London, and Firmin & Sons, London; by Jennens as Charles Jennens until 1912, Jennens & Co until 1924, when they amalgamated with Gaunt & Son. Some other button makers who supplied volunteers were:

S. W. Silver & Co, Clothiers, London
Doughty & Co, 109 St Martins Lane, London
J. W. Reynolds & Co, 50 St Martins Lane, London
Hawkes & Co, 14 Piccadilly, London
Smith & Wright, Birmingham
Hobson & Son(s), London
W. Dowler
I. & B. Pearce & Co, London
William Jones & Co, London

With further research, it would be possible to find the dates during which the firms operated under these titles.

5: Medals and tokens

COLLECTORS OF VOLUNTEER MILITARIA WITH AN INTEREST IN medals have at least three avenues to explore. A few campaign medals were issued to volunteers and to members of yeomanry units; volunteer long service and efficiency awards come next and then, most interesting of all, the private medals issued to volunteers, particularly the many hundreds struck during and just after the Napoleonic Wars.

Britain had not fought a major war since 1855–6 when, with its French and Turkish allies, it showed the world in the Crimea how inefficiently it was officered and administered as compared with the Russians. The volunteer force, although it was born in 1859 against the background of the Crimean fiasco, was designed for home defence and as a protection against a possible threat from France. All the minor colonial wars which took place during the remainder of the nineteenth century were dealt with by the regular forces, and there was always the militia to be called out in emergencies. The war against the Boers in 1899 was a different matter. Vast territories had to be covered, and the determined and effective resistance of the enemy engaged all the available regulars and much of the militia. As always happens in a national emergency, men queued up to serve abroad, men who were equipped and already partly trained—the volunteers. A volunteer company was attached to each regular battalion in the first phase of the war in 1899, consisting of 9,187 men in all. Also, the City of London Imperial Volunteers, the CIV, provided an infantry battalion, a fourteen-gun battery and mounted infantry, altogether 1,600 volunteers coming from units of the Middlesex RV. Mounted infantry proved quite successful against the Boers, who were good horsemen and crack shots with the ability to strike hard and with surprise, and to be away before large groups of infantry could retaliate. Twenty mounted battalions of volunteer Imperial Yeomanry, totalling 10,195 men, were enrolled shortly after the war began, each

MEDALS AND TOKENS

battalion having four companies of approximately 116 men, except No 16, which had three. The companies making up each battalion were numbered from 1 to 79. Later on in the war companies numbered from 80 to 177 were formed, fifteen joining earlier battalions and the remainder forming Imperial Yeomanry battalions given numbers from 21 to 39.

Two campaign medals were awarded for service in South Africa, the Queen's South Africa Medal and the King's South Africa Medal. Twenty-six clasps or bars were authorised for the first, covering the major battles of the campaign, and Cape Colony, Transvaal and Orange Free State. Two clasps were available with the King's Medal, South Africa 1901 and 1902. Both awards were in silver with the monarch's head on one side and the words 'South Africa' and a scene with Britannia waving to troops on the other. The recipient's number, rank, name and regiment were engraved on the rim. In the case of the Imperial Yeomanry, most often the number of the company was given, and with the volunteers attached to the regular infantry, the companies used 'Vol Cy' before or after the title of the regiment. For example, seventy-five men of K (Volunteer Service) company of 1st battalion Queen's Own Cameron Highlanders qualified for campaign medals. According to the medal rolls of that unit, all of them were entitled to the QSA with from one to three clasps, and all but five obtained the KSA as well, with the South Africa 1901 clasp only. As this medal was not normally awarded with only one bar, those issued to the Cameron volunteers would be the basis of an interesting collection.

There is a good deal of scope, too, with the Imperial Yeomanry. It would take a long time to collect a South Africa war medal to each of the 177 companies. CIV South Africa medals do not give the parent corps of the Middlesex RV, but sometimes a medal may be found as a pair to a long service award, which will bear details of the corps to which the recipient belonged.

When the volunteers returned, they were usually provided with some memento by a grateful mayor and corporation. Croydon gave every man an illuminated address, Bideford issued an attractive silver medal cast with an inscription on the reverse showing *Presented to ... 4th V.B. Devon Regt by the town of Bideford for services rendered in South Africa during the war of 1899–1901. June 14th 1901.* The rank, name and company of the recipient were engraved in the space provided.

Medal presented to a volunteer for services in the Boer war

Long-service awards, although they were not given for active service, can nevertheless form an interesting collection. The Volunteer Officers' decoration was issued for twenty years' service, not necessarily continuous, which had to be commissioned, although half of any service in the ranks could be counted towards the qualifying period. It consisted of the letters 'VR' or 'VRI' surmounted by a crown, within an oval wreath of oak-leaves. Suspension was by ribbon from a ring at the top of the medal. This award, although it became known as the Territorial Decoration (the TD) in 1908, has continued through all reigns to the present day, with the appropriate monarch's cypher set within the wreath.

Army Order 85 of 26 May 1894 authorised the issue of the Volunteer Long Service medal to other ranks. To qualify, men had to have been serving volunteers on 1 January 1893, and to have completed twenty consecutive years. As such, it was not easy to obtain, so by 1894 the requirement for continuous service was revoked, and the award was extended to apply to those who had, at any time previously, completed twenty years service. The medal, which was circular and of silver, continued through the reigns of three monarchs to 1930; the oval silver Territorial Efficiency Medal dates from 1908, was modified in 1930 and is still awarded, but now after only twelve years' service.

Imperial Yeomanry long service medals are interesting to collectors, and are now quite expensive and scarce. Authorised in December 1904 by Army Order 211, it was awarded to NCOs and men who had served in the Imperial Yeomanry for ten years on or after 9 November 1904. The oval silver medal has a ring suspension with a yellow ribbon, with an effigy of Edward VII on the obverse and on the reverse the inscription 'Imperial Yeomanry for long service and good conduct'. Edges were impressed with the number, rank, initials, name and regiment of the recipient. Only 1,674 were granted, all between February 1905 and February 1917, to members of fifty-three yeomanry regiments, the most—sixty-five—to the Yorkshire Dragoons and the Royal North Devon, with only three to members of the Glamorganshire yeomanry. To obtain one to each of the fifty-three regiments would be an achievement to keep a collector busy for a lifetime.

All the medals mentioned above are official, standardised awards made by or on behalf of the Crown and follow from the first real campaign medal granted in 1816 to men who fought at Waterloo. Military and naval general service medals for actions during the Napoleonic Wars were first awarded to survivors as late as 1848. This does not mean that medals were not given for other events before 1816, because there were thousands of private medals or medallions issued by the authorities of grateful towns and cities, and by individuals, many of which were struck for presentation to volunteers between 1796 and 1816. Some are fine examples of the medallists' and engravers' craft. Page 72 shows a silver medal presented to the Royal Bristol Volunteers in 1814. On the obverse is shown the arms of the city of Bristol with the motto *Virtute et industria*, outside of which is the inscription *Royal Bristol Volunteers*, with *in danger ready* below. The reverse is inscribed: *Embodied for the maintenance of public order and protection of their fellow citizens, on the threat of invasion by France MDCCXCVII; revived at the renewal of hostilities, MDCCCIII; disbanded when the deliverance of Europe was accomplished by the perseverence and magnanimity of Great Britain and her allies MDCCCXIV G.R. Pro patria*. Many more are listed in Chapter 21 of the second volume of Wheeler and Broadley's *Napoleon and the invasion of England*, published in 1908.

During the 1790s and in the first decade of the nineteenth century, copper penny and halfpenny tokens were in common use in all parts of England. Tokens first appeared in the sixteenth century, issued by tradesmen as a means of overcoming the shortage of small coins available to the poor, and continued in use until new copper coins covering small denominations were issued in 1672. After having been in circulation for over a century, this copper coinage had so deteriorated with constant use—with the added problem that more than half of it was forged—that tokens once again became a substitute for official coins until Matthew Boulton's famous copper cartwheel penny and twopenny pieces were minted at the end of the century. Copper later became so scarce and expensive that the coins were melted down and sold at a profit, resulting in yet another shortage of small change and the reappearance of tokens until the issue of more new coins in 1816.

Tokens, or commercial coins, are quite common and many of those issued during the Napoleonic Wars refer to the volunteer movement. The following is a list of some known examples, but there must be many more which have not yet come to light:

Falmouth Independent Volunteers (1797)
Bristol Volunteers (1798)
Norwich Loyal Military Association (1797)
Blofield Cavalry—5th Troop, Loyal Norfolk Yeomanry (1796)
Norwich Yeomanry Cavalry (1792)
Loyal Suffolk Yeomanry (1794 and 5)

Birmingham Association—presentation of colours (1798)
Warwickshire Yeomanry (1799)
Wiltshire Yeomanry Cavalry (1794)
Portsmouth Green Vols (1791)
Penryn Volunteers (1794)
Somerset Yeomanry Cavalry (1796)
Hoxne Cavalry (1798)
Bath Association (1798)

6: *Weapons*

IN MOST CASES, VOLUNTEERS USED THE SAME WEAPONS AS THE regulars. Sometimes these were of superior quality and finish, and when they came from official WD sources they were often obsolete or superseded models. A few were made exclusively for volunteers. In this chapter, a brief description is given of some of the bayonets, swords and firearms which should enable collectors to identify items of volunteer origin.

Bayonets have been carried as side arms by other ranks throughout the history of volunteer soldiering. With most of the early types in use up until about the middle of the last century, the sole object was to convert the musket into a pike. It took a long time to prepare a muzzle-loading musket for a shot and the first bayonets were merely a dagger with a tapered wooden handle, which plugged into the muzzle of the weapon after it had been discharged. Hence the term 'plug bayonet'. In the heat of an action it was easy to make the mistake of inserting the plug before firing, at best resulting in the loss of the bayonet upon firing the musket. Other disadvantages were met in that the bayonet handle could break off in the barrel, or fall out and become lost if not driven in with sufficient force. The system proved most unsatisfactory and led to the adoption of the ring, and later the socket bayonet, which could be slid over the end of the barrel, leaving the weapon at all times ready for shooting or thrusting.

Weapons suitable for use as both sword and bayonet were devised, but seldom served the two purposes satisfactorily. Indeed, most of them were useless as hand weapons and too long and unwieldy as bayonets. Finally there were bayonets designed to serve as knives and these were quite efficient in both roles.

Napoleonic war volunteers were armed mostly with the flintlock smooth-bore musket then current. It used a triangular socket bayonet 15 to 17in long in a black leather scabbard with a brass top section and a brass tip. Some units used Baker rifles purchased by the commanding officers; the Baker bayonet was a sword variety with a blade 23in long, but it appears from those examples which have survived that volunteers used more elaborate and decorated sword bayonets with this rifle. Some were quite effective as swords, but appear rather too long and flimsy as bayonets. Unfortunately, the only mark on this weapon is that of the maker, and it is therefore difficult to attribute examples to particular units. In 1853 the Enfield socket bayonet was issued to the army and continued in use with later rifles until 1875, being re-bushed to fit the smaller barrels, but volunteers probably used it for a few more years after it became obsolete. Shortly after the Enfield socket was introduced, the 1858-pattern sword bayonet was issued. It differed from earlier sword bayonets in that it had a recurving blade, a shape known as yatagan. Volunteers carried it well into this century, and it was not declared obsolete in the regular army until 1903. Examples are fairly common—including some models which were imported from Germany and privately purchased by volunteers—and were often engraved with the supplier's name—Parker Field & Sons of Holborn, Reilly of Oxford Street etc—sometimes also with the owner's name, but seldom with the unit's designation, unless the weapon was a presentation item.

War Department weapons, as well as both the 1853 and 1858 pattern bayonets issued to volunteer units, were marked and

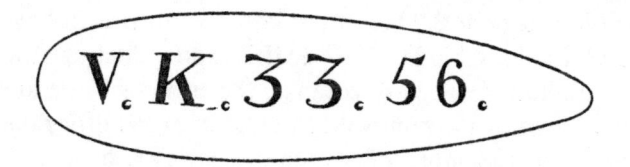

$V_c K_c 3 3_c 5 6_c$

Volunteer bayonet marking to 33 Kent RV c 1860

once the method employed is understood, identification is a simple matter. In September 1860 a War Office letter was circulated to commanding officers of volunteer corps ordering that all government-owned rifles were to be marked by engraving the butt plates with consecutive numbers, the letter 'V', county abbreviations and the number of the corps in its precedence in the county. Bayonets and scabbards were included in this order, so that the markings are to be found engraved on both weapon and scabbard. Thus 'V33K59' may be interpreted as the fifty-ninth bayonet in the series issued to the 33 (Sevenoaks) Kent rifle volunteer corps. On the matching rifle, the marks would appear thus: $\frac{V}{K33}$. The letter 'M' would follow the county letter for a mounted rifle corps, 'A' for volunteer artillery and 'E' for engineer volunteers. This system was used also on the 1876 Martini Henry socket bayonet, which was similar to the 1853 pattern but with a longer blade. On later bayonets, the marking was simplified and most can be interpreted easily—'Ayr Yeo', for example, appeared on bayonets issued to the Ayrshire Yeomanry at the turn of the century.

A change in the style of bayonet blades occurred in 1888 with the introduction of the Lee Metford rifle, a longer weapon than its predecessor, and the double-edged blade was only 12in long, giving an overall length to the rifle and bayonet of 6ft. There were four different models, each with minor modifications resulting from changes in the rifles, until 1903 when the hilt was altered to the beak form used on the longer bayonets of the 1914–18 war. It continued in use in the regular army until 1907, but volunteer and territorial battalions used all these variants for a much longer period. One of the best known British bayonets followed, the long-bladed 1907 SMLE model, first used with a hooked quillon designed to entangle the enemy's blade in combat but altered by the removal of the hook in 1913, presumably because it proved more of a hindrance than a benefit. These were still in use during World War II, but were replaced by a variety of short spiked weapons, which resembled meat skewers and proved very useful for punching holes in tins of condensed milk.

Legislation regarding the ownership of firearms is fairly

straightforward, but a collector wishing to obtain guns would be unwise if he did not first read a copy of the Firearms Act, and then consult his local police as the Act is not at all clear when it comes to defining an antique. In general, a muzzle-loading weapon, whether smooth-bored or rifled, is regarded as an antique provided it has not been obtained for shooting, and there is no reason why they should not be collected. Antiques have been defined as articles which are more than one hundred years old, but there are rifles which were first made during the 1860s which, although they are antiques in this generally accepted sense, are also regarded as firearms. An example is the Snider, and these can be owned only by holders of firearms certificates. These are not easy to obtain, but any citizen without a recent criminal record can only be denied a certificate if it can be shown that ownership of such a weapon would be a danger to the public. The authority granted by the police only covers the weapon or weapons for which initial application is made, and any subsequent guns have to be added to the certificate by means of a variation before they can be acquired. Conversely, a firearm can only be disposed of to a person authorised to receive it, being the holder of a certificate or a special dealer's licence. Collections of firearms must be housed in a secure manner and most authorities will insist upon the weapons being under lock and key and securely chained to a wall when not being cleaned or examined. If a collector wishes to own cartridge revolvers or automatic pistols, then there will be difficulty, and it is hard to show good reason for wishing to have them. Most chief constables will only grant members of shooting clubs firearms certificates for hand guns.

Shotgun certificates enable members of the public to acquire and keep smooth-bore shotguns, and they can be obtained easily from the local police. To overcome the problems experienced in obtaining firearms certificates, a number of dealers are selling even quite modern rifles which have been bored out to a small shotgun calibre, thus converting them technically from rifles to shotguns and no one seems to have found a valid reason why this should not be done. If it is not possible to obtain a firearms certificate, a representative collection of later cartridge rifles, bored out as shotguns, can be bought if they are only required for display purposes. It would still be necessary, of course, to ensure that they are kept secure.

Never consider forming a collection of machine guns. Only those with special Home Office authority may own them and permission is only likely to be given to a museum, an arms manufacturer or a dealer. Machine guns are mentioned here, although they may be regarded by some as being far too modern to have been used by volunteers, because the first British unit to adopt them was the 22nd (Central London Rangers) Middlesex RVC. Lt-Col W. Alt bought two five-barrelled Nordenfelt guns in 1882 from his own funds and displayed them on manoeuvres at Aldershot the following year. Shortly after this the 4 VB Royal Fusiliers and the 26th Middlesex Cyclists' Bn formed machine-gun sections, the latter using a Maxim on a special light carriage hauled behind two cyclists. The British regular army first used Maxim guns in action in the Chitral expedition in 1895.

Amongst the swords used by the early volunteers, the light cavalry sabre of 1796 which continued in use until 1829 is the most common. It had a curved blade with a single cutting edge; the back of the hilt had circular lobes on each side which extended across the grip and were held together by a rivet. The knuckle guard was stirrup-shaped and there were two short guards, lying parallel to the blade, which fitted outside the scabbard. It is on one of these lobes, and at the top of the scabbard, that the unit's designation was stamped or engraved. They are not always easy to interpret, but often include the letters 'YC' to denote Yeomanry Cavalry. Examples of superior quality, issued to officers as presentation pieces, and sometimes bearing engraved inscriptions, are highly desirable and much sought after by collectors. Officers of the Napoleonic volunteer infantry regiments seem to have adopted as standard a sword with a straight gently tapering blade about 30in long with fullers

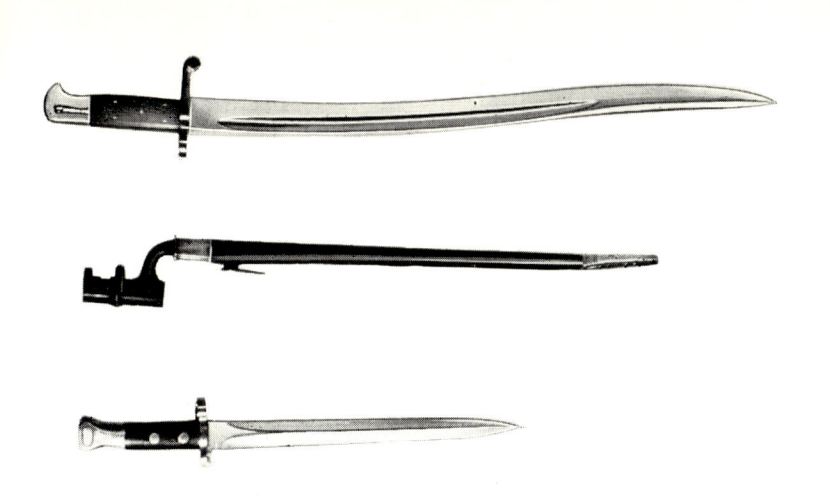

Page 71 (above) *Volunteer Bayonets. 1856 pattern Enfield sword bayonet. 1853 pattern Enfield socket. 1888 pattern Lee-Metford;* (below) *officers' full dress pouch and crossbelt of the 3rd Kent Artillery Volunteers pre-1901 and, beneath, leather pouch and belt of the Fifeshire Volunteer Rifles c 1880*

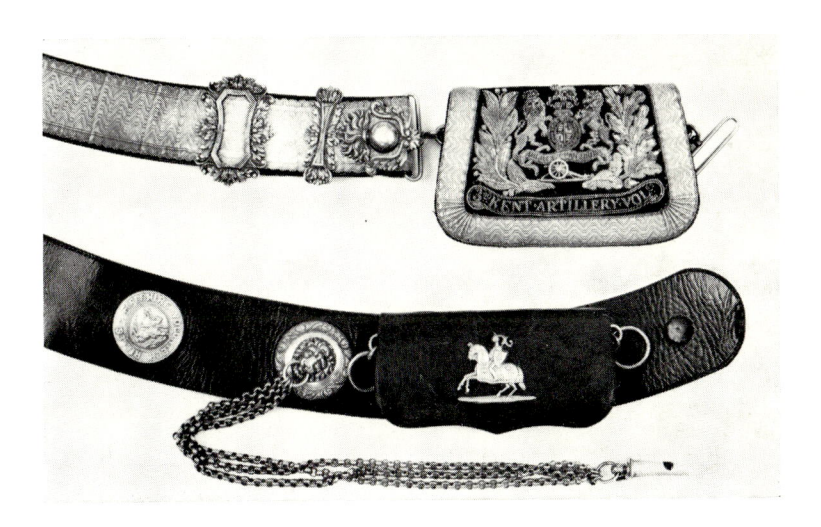

Page 72   *1, Silver medal for service in the Royal Bristol Volunteers issued in 1819. Volunteer buttons: 2, Robin Hood Rifles; 3, London Scottish Volunteers; 4, General issue, Rifle Volunteers; 5, Cheshire Rifle Volunteers; 6, Berkshire Rifle Volunteers*

grooved down each side of its full length. The grip was of ebony, often with an oval silver inset on which was engraved the name of the corps, with a single gilded brass knuckle guard, D-shaped at the pommel with a right-angle bend towards the grip where it joined the blade.

Rifle volunteer officers from 1859 generally bought nickel-plated swords of the pattern used by rifle regiments. The steel guard, known as the 'gothic' type and of a shape introduced in 1822 and adopted by rifle regiments in 1834, contained a stringed bugle emblem in place of the royal cypher. The county precedence number was sometimes engraved under the bugle, and the rifle volunteer title on one side of the blade. Infantry volunteer other ranks were not permitted to carry swords, although some of them did so in the early days, but all yeomanry ranks carried them. Officers used the 1822 pattern light cavalry weapon with the three-bar hilt, often these were engraved with the royal cypher on one side of the blade and the title of the regiment on the other. Hilts were either silver- or nickel-plated. In 1853, a new cavalry sword was introduced with riveted leather grips and a three-bar steel hilt for other ranks. It was followed by three basic types, including some scarce and complicated variations, all with sheet steel guards. The first was the 1864 pattern, characterised by four holes pierced in the form of a Maltese cross on the guard. Next came the heavy, unwieldy pattern of 1899, followed by the last design in 1908 which resulted in a weapon for thrusting only and was the best sword ever given to British troops. All these varieties can be found with yeomanry markings.

By the time the first volunteer corps were formed in 1859, the flintlock principle of ignition had been replaced by the percussion system. The rifle was still loaded from the muzzle, but the charge was fired, not by a spark caused by flint striking on a steel, but by the flame caused as a result of striking fulminate of mercury held in a small copper cap placed over a nipple underneath a hammer. Introduced in 1853, the Enfield rifle taking a bullet of 0·577 diameter became the symbol of the

E

patriotism of rifle volunteers, and the last muzzle-loading rifle of the British army.

Until 1861, volunteer corps could either draw arms from government stores or purchase their own, provided that the rifles would accept the standard ammunition and conformed to the same basic design as the military rifle. In February, a War Office letter ordered that all corps should be issued with arms supplied by the War Office, but many members had already bought rifles privately and continued to use them. There was a great deal of bad feeling created by the issue of inferior Enfield rifles, particularly when the War Office said that it would buy, at government rates, any privately-owned rifles to be taken into ordnance stores. As competition shooting was such an important aspect of the movement, it was necessary for volunteers to have good weapons.

D. Bailey, in an article in *Guns Review*, categorised volunteer rifles into three classes: those purchased and used by non-commissioned volunteers; those made for officers; and rifles given as prizes at rifle meetings, amongst which were 0·451 calibre arms such as the Whitworth, Kerr and Turner. These rifles, which date mostly from 1859 and 1860, do not have the 'Tower' or 'Enfield' marks on the locks and are of superior quality and finish, with engraved furniture and chequered stocks. They are still to be found in sufficient numbers to make a good collection possible. Volunteers using regulation Enfield arms had to make do with second-class weapons, which bore a large '2' stamped on the right-hand side of the butt. Regimental markings, as described earlier, will also be found on the second-quality Enfields.

Many experiments were made to perfect a system of breech loading at about this time. Most methods tried failed to overcome the problem of sealing in the gases at the breech, and it was not until the metal cartridge was perfected that a full-scale issue of breech-loading rifles became possible. In Britain, this was achieved in 1866 by converting existing Enfield muzzle-loaders to the Snider system, in which a hinged block kept the cartridge in position. A firing pin was held obliquely in the block and was struck by the hammer in the same manner as the percussion cap in the muzzle-loader. One of the transitional breech-loaders did survive amongst the yeomanry in the form of the Westley Richards 'monkey-tail', which was adopted as the standard cavalry arm in 1861, withdrawn on the introduction of the Snider five years later, and remained with the yeomanry for many years.

Sniders were replaced by the Martini-Henry, adopted in 1871; which in turn gave way to the magazine rifle of a similar type to that in use until the fairly recent adoption of the self-loading semi-automatic weapon of today.

7: *Rifle Volunteer Corps and their successors*

COLLECTORS WILL FIND MANY QUESTIONS WHICH ARE DIFFICULT to answer. What, for example, is the significance of the number 19 on the small cap badge of the Berkshire Rifle Volunteers? What is the connection between the 20th Middlesex RV and the 10th Bn R Fusiliers? Cambridgeshire was numbered 47 in the volunteer precedence; what does the number 9 mean on the buttons of an old Cambridgeshire tunic? This chapter will enable collectors to answer questions like this and to trace the outline of the history of a particular unit from its formation until it became part of the Territorial Force in 1908. In some cases, details are given up to 1937.

Basically, the information comes from the *Monthly Army Lists* for February 1861, January 1872, the lists of units given in Major Walter's *The Volunteer Force, history and manual*, published in 1881, and in the *Territorial Year Book* for 1909. By 1861, the majority of individual volunteer corps had been formed in towns and villages, most of them had been grouped into administrative battalions by 1872; by 1881, the small sub-units had lost their individual identity just prior to becoming volunteer battalions of the re-organised regular infantry regiments with county affiliations, which were formed from the numbered regiments. In 1908, the volunteer battalions were re-numbered to form the Territorial Force.

The complete story from 1859 to the present day is both highly complicated and beyond the scope of this book, but the collector should find sufficient information in the following pages to assist with the dating of volunteer items.

Each county is listed alphabetically, the number in parentheses following each heading being the number of precedence of the county and of all the separate corps and administrative battalions listed beneath. The corps numbers at the left of the page relate to the precedence of each separate corps within the county.

RIFLE VOLUNTEER CORPS

An asterisk against a title shows that, although the title of the corps is included in the *Army List*, no officers were listed in the edition used. Numbers in parentheses after the individual unit titles show the numbers of the administrative battalions into which they were grouped.

After the county titles, the number (i) is followed by the colour of the uniform in 1880 and (ii) by the colour of the facings of the uniform in 1880. After the list of all the corps in each county in 1861, 1872 and 1880, there follows notes of their subsequent histories up to, at least, 1908. The number in parentheses after each volunteer battalion title denotes the post-1881 order of precedence.

As an example, Hertfordshire, being the fifty-first county in order of precedence in 1859, was 51 Hertfordshire VRC. It had a corps in Hertford which was the 1st Hertfordshire VRC, being a part of the 2nd Administrative Bn Hertfordshire VRC by 1861. In 1872, the 22nd Essex VRC was united with the 2nd Hertfordshire Adm Bn, to become—by 1880—the 1st (Herts and Essex) Hertfordshire VRC, wearing a grey uniform with scarlet facings. The re-styling of 1881 created the 1st Volunteer Bn Bedfordshire Regt, which became the 1st Bn Hertfordshire Regt in 1908. This gives a collector interested in Hertford a goal of more than seven different badges and buttons to cover the period from 1859 to 1908.

RIFLE VOLUNTEER ORDER OF PRECEDENCE—1872

1 Devonshire
2 Middlesex
3 Lancashire
4 Surrey
5 Pembrokeshire
6 Derbyshire
7 Oxfordshire
8 Cheshire
9 Wiltshire
10 Sussex
11 Edinburgh (City)
12 Essex
13 Northumberland
14 Renfrewshire
15 Northamptonshire
16 Dorsetshire
17 Norfolk
18 Staffordshire
19 Berkshire
20 Gloucestershire
21 Brecknockshire
22 Suffolk
23 Stirlingshire
24 Buckinghamshire
25 Lanarkshire
26 Kent
27 Glamorgan
28 Nottinghamshire
29 Merionethshire
30 Yorkshire
   (W Riding)
31 Leicestershire
32 Midlothian
33 Aberdeenshire
34 Roxburgh
35 Cinque Ports
36 Monmouthshire
37 Cornwall
38 Ross-shire
39 Worcestershire
40 Inverness-shire
41 Warwickshire
42 Lincolnshire
43 Denbighshire
44 Hampshire
45 Somersetshire
46 Forfar
47 Cambridgeshire
48 Shropshire
49 London
50 Yorkshire
   (E Riding)
51 Hertfordshire
52 Perthshire
53 Berwickshire
54 Sutherland
55 Kincardineshire
56 Haverfordwest
57 Haddington
58 Isle of Wight
59 Ayrshire
60 Dumfries
61 Elgin
62 Argyll
63 Cardiganshire
64 Durham
65 Wigtown
66 Buteshire
67 Yorkshire
   (N Riding)
68 Cumberland
69 Herefordshire
70 Dumbarton
71 Huntingdon
72 Carnarvonshire
73 Montgomeryshire
74 Orkney
75 Carmarthen
76 Caithness
77 Kirkcudbright
78 Westmorland
79 Fifeshire
80 Bedfordshire
81 Newcastle-on-Tyne
82 Linlithgowshire
83 Selkirkshire
84 Banffshire
85 Radnorshire
86 Flintshire
87 Berwick-on-Tweed
88 Clackmannan
89 Tower Hamlets
90 Nairn
91 Peeblesshire
92 Isle of Man
93 Kinross-shire
94 Anglesey

1861	1872	1880
Aberdeenshire (33)	(i) Green (ii) Scarlet	
1—Aberdeen	Aberdeen	Aberdeen
2—Tarves	Methlie (2)	Aberdeen
3—Cluny (1)	Cluny (1)	(The Buchan) Old Deer
4—Alford (1)	Alford (1)	Aberdeen
5—New Deer	New Deer (3)	—
6—Ellon	Ellon (2)	—
7—Huntly (1)	Huntly (1)	—
8—Echt (1)	Echt (1)	—
9—Peterhead	Peterhead (3)	—
10—Inverury (1)	Inverury (1)	—
11—Kildrummy (1)	Kildrummy (1)	—
12—Old Aberdeen	Newmachar (2)	—
13—Turriff	Turriff (2)	—
14—Tarland	Tarland (1)	—
15—Fyvie	Fyvie (2)	—
16—Meldrum	Meldrum (2)	—
17—Old Deer	Old Deer (3)	—
18—Aberdeen*	Tarves (2)	—
19—	Insch (1)	—
20—	— Longside (3)	—
21—	— (Marquis of Huntly's Highland) Arboyne (1)	—
22—	— Auchmull (1)	—
23—	— Torphins (1)	—
24—	— St Fergus (3)	—
25—	— New Pitsligo (3)	—

1st RV (1859) became 1 VB Gordon Highlanders (141) in 1883, 4 Bn Gordon Highlanders in 1908.

2nd and 3rd RV became 2nd (142) and 3rd (143) in 1883, amalgamated in 1908 and shortly after became 5 (Buchan and Formartin) Bn Gordon Highlanders.

(*See Banff for 6 Bn and Kincardine for 7 Bn Gordon Highlanders.*)

Anglesey (94)		
1—Almwch	—	—
2—Aberffraw*	—	—
3—Menai Bridge*	—	—

From 1908 recruits from Anglesey joined 6 Bn Royal Welsh Fusiliers (formerly 3 VB (208)).

RIFLE VOLUNTEER CORPS

1861	1872	1880
Argyll (62)	(i) Scarlet (ii) Yellow	
1— —	Dunoon	Argyllshire Highland RV
2—Inverary	Inverary	—
3—Campbeltown†	Campbeltown	—
4— —	—	—
5—Mull	—	—
6—Melford	—	—
7—Dunoon	Dunoon	—
8—Cowal	Glendaruel	—
9—Glenorchy	—	—
10—Tayvollick	—	—
11—Oban	—	—
12— —	—	—
13— —	—	—
14— —	Kilmartin	—

† *United with 3 Perthshire RVC before 1861.*
Became 5 VB Argyll and Sutherland Highlanders (194), and 8 Bn Argyll and Sutherland Highlanders in 1908.

Ayrshire (59)	(i) Scarlet (ii) Blue	
1—Kilmarnock (1)	Kilmarnock (1)	Ayr
2—Ervine (1)	Ervine (1)	Kilmarnock
3—Ayr (1)	Ayr (1)	—
4—Largs (1)	Largs (1)	—
5—Maybole (1)	Maybole (1)	—
6—Beith (1)	Beith (1)	—
7—Saltcoats and Stevenston (1)	Saltcoats (1)	—
8—Colmonell (1)	Colmonell (1)	—
9—Kilmarnock (1)	Kilmarnock (1)	—
10—Girvan (1)	Girvan (1)	—
11—Dalry (1)	Dalry (1)	—
12—Cumnock (1)*	Cumnock (1)	—
13— —	Sorn (1)	—
14— —	Ayr (1)	—

1 RV (1859) became 1 VB Royal Scots Fusiliers (190), and 4 Bn Royal Scots Fusiliers in 1908.
2 RV became 2 VB (191), and 5 Bn Royal Scots Fusiliers.

RIFLE VOLUNTEER CORPS

1861	1872	1880
Banffshire (84)	(i) Grey (ii) Black	
1—Macduff*	Banff (1)	Keith
2—Banff	Aberlour (1)	—
3—Aberlour	Keith (1)	—
4—Keith	—	—
5— —	Buckie (1)	—
6— —	Minmore, Glenlivel (1)	—
7— —	Dufftown (1)	—

Became 4 VB Gordon Highlanders (144) and 6 VB Gordon Highlanders (215), grouped into 6 (Banff and Donside) Bn Gordon Highlanders in 1908.

Bedfordshire (8)	(i) Scarlet (ii) Yellow	
1—Bedford (1)	Bedford (1)	Bedford
2—Toddington (1)	Toddington (1)	—
3— —	—	—
4—Dunstable (1)	Dunstable (1)	—
5 Ampthill (1)	Ampthill (1)	—
6—Luton (1)	Luton (1)	—
7—Biggleswade (1)	Shefford (1)	—
8—Woburn (1)	Woburn (1)	—
9— —	Bedford (1)	—

Became 3 VB (213) and 4 VB (207) Bedfordshire Regiment, grouped into 5 Bn Bedfordshire Regt in 1908.

1 (181) and 2 (182) VB's Bedfordshire Regt were converted to 2 Herts Batts RFA, 4 East Anglian Amb Coy, and Hertfordshire Bn in 1908.

Berkshire (19)	(i) Scarlet (ii) Lincoln Green	
1—Reading (1)	Reading (1)	Reading
2—Windsor (1)	Windsor (1)	—
3—Newbury (1)	Newbury (1)	—
4—Abingdon (1)	Abingdon (1)	—
5—Maidenhead (1)	Maidenhead (1)	—
6—Wokingham (1)	—	—
7—Sandhurst (1)	Sandhurst (1)	—
8—Farringdon	Farringdon (1)	—
9—Wantage	Wantage (1)	—
10— —	Winkfield (1)	—
11— —	Wallingford (1)	—
12— —	Windsor Gt Park (1)	—

Became 1 VB R Berkshire Regt (99) and 4 Bn R Berkshire Regt in 1908.

RIFLE VOLUNTEER CORPS

1861	1872	1880
Berwickshire (53)	(i) Scarlet (ii) Scarlet	
1—Dunse	Dunse (1)	Coldstream
2—Coldstream	Coldstream (1)	—
3—Ayton	Ayton (1)	—
4—Greenlaw	Greenlaw (1)	—
5—Lauderdale	Lauderdale (1)	—
6— —	Earlstoun (1)	—
7— —	Chirnside (1)	—

Became 2 VB KOSB (185) and in 1908 part of 4 (The Border) Bn KOSB.

Berwick-on-Tweed (87)
1—Berwick   —   —

By 1872 this corps was united with 1 A Bn Northumberland RV, prior to which it became part of the Berwickshire RV.

Brecknockshire (21)	(i) Grey (ii) Black	
1—Brecon (1)	Brecon (1)	Brecon
2—Brynmawr (1)	Brynmawr (1)	—
3—Crickhowell (1)	Crickhowell (1)	—
4—Hay (1)	Hay (1)	—
5—Builth (1)	Builth (1)	—

Became 1 (Brecknockshire) VB S Wales Borderers (103), and Brecknockshire Bn SWB in 1908.

RIFLE VOLUNTEER CORPS

1861	1872	1880
Buckinghamshire (24)	1st (i) Grey (ii) Scarlet	
	2nd (i) Grey (ii) Lt Blue	
1—Gt Marlow	Gt Marlow (1)	Gt Marlow
2—Wycombe	High Wycombe (1)	Eton College
3—Buckingham	Buckingham (1)	—
4—Aylesbury	Aylesbury (1)	—
5—Slough	Slough (1)	—
6—Newport Pagnell	—	—
7—Winslow*	—	—
8— —	Eton (1)	—

1 Bucks VRC (107) became Bucks Bn Oxfordshire and Buckinghamshire Light Inf in 1908.
2 Bucks VRC (108) became Eton College OTC.

Buteshire (66)
1—Rothesay   Rothesay   —

United to 1 A Bn Renfrewshire RV before 1872, did not form a territorial unit in 1908.

Caithness (76)
1—Thurso   Thurso
2— —   Wick

By 1872 both corps had been united with 1 A Bn Sutherland RV, becoming 5 Bn (Sutherland and Caithness) Seaforth Highlanders in 1908.

Cambridgeshire (47)	1st (i) Scarlet (ii) Blue	
	2nd (i) Grey (ii) Grey	
1—Cambridge	Cambridge (2)	Cambridge
2—Wisbech (1)	Wisbech (1)	Cambridge Univ
3—Cambridge Univ	Cambridge Univ	—
4—Whittlesey (1)	Whittlesey (1)	—
5—March (1)	March (1)	—
6—Ely (1)	Ely (1)	—
7—Upwell (1)	Upwell (1)	—
8—Cambridge	—	—
9—Newmarket*	—	—

Became 3 (Cambs) VB Suffolk Regt (172) and 1 Bn Cambridgeshire Regt in 1908.
Cambridge University VRC (173) became Cambridge University OTC.

	1861	1872	1880
Cardiganshire (63)			
1—Aberystwyth	—	—	
2—Aberystwyth	Cardigan	—	
3—Aberbank	—	—	
4—Cardigan	—	—	

By 1872, the Cardigan corps had amalgamated with 1 A Bn Pembrokeshire RV.
Territorial recruiting from 1908 was into 4 Bn Welsh Regt.

Carmarthenshire (75)		
1—Llandilo	Llandilo (1)	—
2—Carmarthen	Carmarthen (1)	—
3—Llandovery	Llandovery (1)	—
4—Llansawel	—	—
5—Llanelly*	Llanelly (1)	—
6—	Carmarthen (1)	—

Amalgamated into 1 (Pembrokeshire) VB Welsh Regt (62).
Territorial recruiting from 1908 was into 4 Bn Welsh Regt.

Carnarvonshire (72)		
1—Carnarvon (1)	Penrhyn (1)	—
2—Carnarvon (1)	Penrhyn (1)	—
3—Carnarvon (1)	Carnarvon (1)	—
4—Tremadoc (1)	Tremadoc (1)	—
5—Pwllheli (1)	Pwllheli (1)	—
6—Bangor (1)	—	—
7—Conway (1)	—	—

Amalgamated into 3 VB R Welsh Fusiliers (208), which became 6 (Carnarvonshire and Anglesey) Bn RWF in 1908.

Cheshire (8)   1st (i) Grey   (ii) Scarlet   2nd (i) Grey   (ii) Blue
3rd (i) Scarlet   (ii) White   4th (i) Scarlet   (ii) Buff
5th (i) Grey   (ii) Scarlet

1—Birkenhead (1)	Birkenhead (1)	Birkenhead
2—Oxton (1)	Oxton (1)	Chester
3—Wallasey (1)	Wallasey (1)	Knutsford
4—Bebbington (1)	Bebbington (1)	Stockport
5—Congleton (4)	Congleton (5)	Congleton
6—Chester (2)	(The Earl of Chester's) (2)	—
7—Runcorn (2)	Runcorn (2)	—
8—Macclesfield (4)	Macclesfield (5)	—
9—Mottram (4)	—	—

RIFLE VOLUNTEER CORPS

1861	1872	1880
10—	—	—
11—Neston (1)	Neston (1)	—
12 Altrincham (3)	Altrincham (3)	—
13—Dukinfield (4)	Newton Moor (4)	—
14—Hooton (1)	Hooton (1)	—
15—Knutsford (3)	Knutsford (3)	—
16—Sandbach (4)	Sandbach (5)	—
17—Stockport (4)	Stockport (4)	—
18—Stockport (4)	Stockport (4)	—
19—Stockport (4)	Stockport (4)	—
20—Stockport (4)	Stockport (4)	—
21—Stockport (4)	Stockport (4)	—
22—Northwich (3)	Northwich (3)	—
23—Weaverham (2)	Weaverham (2)	—
24—Frodsham (2)	Frodsham (2)	—
25—Timperley (3)	—	—
26—Northenden (3)	Cheadle (3)	—
27—Wilmslow (4)	Wilmslow (5)	—
28—Sale Moor (3)	Sale Moor (4)	—
29—Stockport (4)	Stockport (4)	—
30—Tranmere (1)	Tranmere (1)	—
31—Hyde (4)	Hyde (4)	—
32—Lymm (3)	Lymm (3)	—
33—Nantwich (5)	Nantwich (5)	—
34—	—	—
35—	Bromborough (1)	—
36—	Crewe (5)	—

1 RVC became 1 VB Cheshire Regt (67) and 4 Bn Cheshire Regt in 1908.
2 and 3 RVC became 2 and 3 VB Cheshire (68 and 69) and amalgamated to form 5 Bn (Earl of Chester's) Cheshire Regt in 1908.
4 RVC became 4 VB Cheshire (70) and 6 Bn Cheshire in 1908.
5 RVC became 5 VB Cheshire (71) and 7 Bn Cheshire in 1908.

Cinque Ports (35)	(i) Grey (ii) Blue	
1—Hastings (1)	Hastings (1)	Hastings
2—Ramsgate (1)	Ramsgate (2)	—
3—Rye (1)	Tenderden	—
4—Hythe (1)	Hythe (2)	—
5—Folkestone (1)	Folkestone (2)	—
6—Deal (1)	—	—
7—Margate (1)	Margate (7)	—
8—Dover (1)	Dover (8)	—

RIFLE VOLUNTEER CORPS

	1861	1872	1880
9—	—	Rye (1)	—
10—	—	New Romney (2)	—

3 Corps (Tenderden) was united with Kent RVC before 1872.
Became 1 Cinque Ports VRC (146) and 5 (Cinque Ports) Bn R Sussex Regt in 1908.

Clackmannon (88)	(i) Scarlet (ii) Blue	
1—Alloa	Alloa (1)	Alloa
2—Tillicoultry	Tillicoultry (1)	—

Recruiting from Clackmannon and Kinross after 1881 was into 7 VB Argyll and Sutherland Highlanders (217), and after 1908 into 7 Bn Argyll and Sutherland Highlanders.

Cornwall (37)	1st (i) Grey (ii) Scarlet 2nd (i) Scarlet (ii) White	
1—Penzance (1)	Penzance (1)	Penzance
2—Camborne (1)	Camborne (1)	(Duke of Cornwall's) Bodmin
3—Falmouth (1)	Falmouth (1)	—
4—Liskeard (2)	Liskeard (2)	—
5—Callington (2)	Callington (2)	—
6—Launceston (2)	Launceston (2)	—
7—Helston (1)	Helston (1)	—
8— —	—	—
9—St Austell (2)	St Austell (2)	
10—Bodmin (2)	Bodmin (2)	—
11—Truro (1)	Truro (1)	—
12—Truro (1)	Truro (1)	—
13—Wadebridge (2)	Wadebridge (2)	—
14— —	—	—
15—Hayle (1)	Hayle (1)	—
16—St Columb	St Columb (2)	—
17—Redruth (1)	Redruth (1)	—
18—Helston	Trelowarren (1)	—
19—Camelford	Camelford (2)	—
20—St Just in Penwith	St Just in Penwith	—
21—Penryn	Penryn (1)	—
22— —	Saltash (2)	—

Became 1 VB DCLI (150) and 2 VB DCLI (151), and 4 Bn DCLI and 5 Bn DCLI after 1908.

1861	1872	1880
Cumberland (68)	(i) Scarlet (ii) White	
1—Carlisle (1)	Carlisle (1)	Keswick
2—Whitehaven (1)	Whitehaven (1)	—
3—Keswick (1)	Keswick (1)	—
4—Brampton (1)	Brampton (1)	—
5—Penrith (1)	Penrith (1)	—
6—Alston (1)	Alston (1)	—
7—Workington (1)	Workington (1)	—
8—Cockermouth (1)	Cockermouth (1)	—
9—Whitehaven (1)	—	—
10—Egremont (1)	Egremont (1)	—
11—Wigton (1)	Wigton (1)	—

Became 1 VB Lancashire Regt (39), changed to 2 (Westmorland) VB Border Regt (210) and to 4 (Cumberland and Westmorland) Bn Border Regt in 1908.

Also 3 (Cumberland) VB Border Regt (204) and 5 (Cumberland) Bn Border Regt in 1908.

Denbighshire (43)	(i) Scarlet (ii) Blue	
1—Wrexham (1)	Wrexham (1)	Ruabon
2—Ruabon (1)	Ruabon (1)	—
3—Denbigh (1)	Denbigh (1)	—
4—Gresford (1)*	Gresford (1)	—
5—Giversyllt*	Gwersyllt (1)	—
6—Ruthin*	Ruthin (1)	—
7— —	Chirk (1)	—
8— —	—	—
9— —	Llangollen (1)	—

Became 1 VB R Welsh Fusiliers (161), and 4 (Denbighshire) Bn R Welsh Fusiliers, with one company transferring to 5 Bn, in 1908.

RIFLE VOLUNTEER CORPS

1861	1872	1880
Derbyshire (6)	1st (i) Scarlet (ii) White	
	2nd (i) Scarlet (ii) Blue	
1—Derby (1)	Derby (1)	Derby
2—Sudbury (2)	—	Bakewell
3—Chesterfield (3)	Chesterfield (3)	—
4—Derby (1)	Derby (1)	—
5—Derby (1)	Derby (1)	—
6—Buxton (3)	—	—
7—Chapel-en-le-Frith (3)	Chapel-en-le-Frith (3)	—
8—Ashbourne (2)	Ashbourne (3)	—
9—Bakewell (3)	Bakewell (3)	—
10—Wirlesworth (2)	Wirksworth (3)	—
11—Matlock (3)	Matlock (3)	—
12—Butterley (1)	Butterley (1)	—
13—Belper (1)	Belper (1)	—
14—	—	—
15—Derby (1)	Derby (1)	—
16—Ilkeston	—	—
17—Clay Cross*	Clay Cross (3)	—
18— —	Whaleybridge (3)	—
19— —	(Elvaston) Derby (1)	—
20— —	(Trent) Long Eaton (1)	—

Became 1 VB Notts and Derby Regt (63), reformed as 5 Bn Notts and Derby Regt in 1908, and 2 VB Notts and Derby Regt (64), reformed as 6 Bn Notts and Derby Regt.

Devonshire (1)	1st (i) Green (ii) Black	2nd (i) Green (ii) Scarlet
	3rd (i) Grey (ii) Green	4th (i) Scarlet (ii) White
	5th (i) Scarlet (ii) Green	
1—(Exeter and S Devon) Exeter	(Exeter and S Devon) Exeter	(Exeter and S Devon) Exeter
2—Plymouth (3)	Plymouth (2)	(Prince of Wales') Plymouth
3—Devonport (3)	Devonport (2)	Exeter
4—Ilfracombe (4)	Ilfracombe (3)	Barnstaple
5—Upper Culm Vale (2)	Cullompton (1)	Newton Abbot
6—Barnstaple (4)	Barnstaple (3)	—
7—	—	—
8—Buckerell (2)	Buckerell (1)	—
9—Ashburton (5)	Ashburton (4)	—
10—Newton Abbot (5)	Newton Abbot (4)	—
11—Bampton (2)	Bampton (1)	—
12—	—	—
13—Honiton (2)	Honiton (1)	—

Page 89 (top) *OR's white-metal helmet plate;* (bottom) *OR's white-metal helmet plate centre of 2nd VB KOSB 1881–1901;* (left) *glengarry badge of the Berkshire RV 1873–81;* (right) *shako badge of the 24th Middlesex RV, the Post Office Rifles post 1880*

Page 90 *Cap badges post 1901: 1, London Scottish; 2, Tyneside Scottish; 3, 2 VB Hampshire Regt; 4, R Gloucestershire Hussars Imperial Yeomanry; 5, 12 Bn The London Regt; 6, Artists Rifles; 7, Artists*

1861	1872	1880
14—Tiverton (2)	Tiverton (1)	—
15— —	—	—
16—Stonehouse (3)	Stonehouse (2)	—
17—Totness (5)	Totnes (4)	—
18—Hatherleigh (4)	Hatherleigh (3)	—
19— —	—	—
20—Broadhembury (2)	Broadhembury (1)	—
21—Bideford (4)	Bideford (3)	—
22—Tavistock (3)	Tavistock (2)	—
23—Chudleigh (5)	Chudleigh (4)	—
24— —	—	— BICTON
25—Ottery St Mary (2)	Ottery St Mary (1)	—
26—Kingsbridge (5)	Kingsbridge (4)	—
27—Colyton*	Colyton (1)	—
28— —	South Brent (4)	—

Became 1 VB (1) and 3 VB (3) Devonshire Regt, 4 Bn Devonshire Regt in 1908.
2 VB (2) and 5 VB (5) Devonshire Regt, 5 Bn Devonshire Regt in 1908.
4 VB Devonshire Regt (4), 6 Bn Devonshire Regt in 1908.
Recruits from the south coast of Devon also joined 7 (Cyclist) Bn Devonshire Regt from 1908.

Dorset (16)	(i) Green (ii) Scarlet	
1—Bridport	Bridport (1)	Dorchester
2—Wareham (1)	Wareham (1)	—
3—Dorchester (1)	Dorchester (1)	—
4—Poole (1)	Poole (1)	—
5—Weymouth (1)	Weymouth (1)	—
6—Wimborne (1)	Wimborne (1)	—
7—Sherborne (1)	Sherborne (1)	—
8—Blandford (1)	Blandford (1)	—
9—Shaftesbury (1)	Shaftesbury (1)	—
10—Sturminster (1)	Sturminster (1)	—
11—Gillingham	Gillingham (1)	—
12—Stallbridge	Stallbridge (1)	—

Became 1 VB Dorsetshire Regt (89) and 4 Bn Dorsetshire Regt in 1908.

Dumbarton (70)	(i) Green (ii) Scarlet	
1—Row (1)	Row (1)	Helensburgh
2—E Kirkpatrick (1)	Maryhill (1)	—
3—Bonhill (1)	Bonhill (1)	—
4—Jameston (1)	Jameston (1)	—
5—Alexandria (1)	Alexandria (1)	—

F

1861	1872	1880
6—Dumbarton (1)	Dumbarton (1)	—
7—Cardross (1)	Cardross (1)	—
8—Gareloch (1)	—	—
9—Luss (1)	Luss (1)	—
10—Kirkintilloch (1)	Kirkintilloch (1)	—
11—Cumbernauld (1)	Cumbernauld (1)	—
12—	—	—
13—	Milngaire (1)	—

Remained 1 Dumbarton RV (206) and became 9 (Dunbartonshire) Bn Argyll and Sutherland Highlanders in 1908.

Dumfries (6)	(i) Scarlet (ii) Yellow	
1—Dumfries	Dumfries (1)	Dumfries
2—Thornhill	Thornhill (1)	—
3—Sanquhar	Sanquhar (1)	—
4—Penpont	Penpont (1)	—
5—Annan	Annan (1)	—
6—Moffat	Moffat (1)	—
7—Langholm	Langholm (1)	—
8—Lockerbie	Lockerbie (1)	—
9—Lochmaben*	Lochmaben (1)	—

The 1 Dumfries Mounted RVC was attached to the corps by 1880. It became 3 VB KOSB (192) and Galloway VRC (200), and was formed into 5 (Dumfries and Galloway) Bn KOSB in 1908.

Durham (64) 1st (i) Scarlet (ii) White 2nd (i) Green (ii) Scarlet
             3rd (i) Scarlet (ii) Blue   4th (i) Green (ii) Scarlet
             6th (i) Scarlet (ii) Dk Green

1—Stockton-on-Tees	Stockton-on-Tees (4)	(Durham & N Riding of York)
2—	—	Bishop Auckland
3—Sunderland	(The Sunderland)	(The Sunderland)
4—Bishop Auckland (2)	Bishop Auckland (2)	Chester-le-Street
5—	—	—
6—South Shields	Tyne Docks (3)	Gateshead
7—Durham (1)	Durham (1)	—
8—Gateshead	Gateshead (3)	—
9—Blaydon	Blaydon Burn (3)	—
10—Beamish (1)	Beamish (1)	—
11—Chester-le-Street (1)	Chester-le-Street (1)	—
12—Middleton Teesdale (2)	Middleton Teesdale (2)	—

1861	1872	1880
13—Birtley (1)	Birtley (1)	—
14—Felling	Felling (1)	—
15—Darlington (2)	Darlington (4)	—
16—Castle Eden	Castle Eden (4)	—
17—Wolsingham (2)	—	—
18—Shotley Bridge	—	—
19—Hartlepool*	Hartlepool (4)	—
20—Stanhope*	Stanhope (2)	—
21— —	Barnard Castle (2)	—

Became 1 VB DLI (195), and 5 Bn DLI in 1908.
2 VB DLI (196), and 6 Bn DLI in 1908.
4 VB DLI (198), and 8 Bn DLI in 1908.
5 VB DLI (199), and 9 Bn DLI in 1908.
Recruits from Durham also joined the Northern Cyclists Bn.

Edinburgh (City) (11)	1st (i) Grey (ii) Grey	
	2nd (i) Scarlet (ii) Blue	
1—(City of Edinburgh)	(The Queen's City of Edinburgh RV Brigade)	(The Queen's City of Edinburgh RV Brigade)
2— —	—	—
3— —	†	—

† *By 1872 the 3rd Edinburgh was attached to the 1st Brigade, but was still regarded as a separate corps, known as 2nd Edinburgh, in 1880.*

The 1st became The Queen's in 1865 and was first formed in August 1859.

Became The Queen's RV Brigade The Royal Scots (Lothian Regt) in 1888 (76) and 4 and 5 Bn R Scots (Queen's Edinburgh Rifles) in 1908.

4 VB R Scots (77) became 4 Bn R Scots in 1908.

9 VB R Scots (78) formed in August 1900, became 9 (Highlanders) Bn R Scots in 1908.

	1861	1872	1880
Elgin (61)		(i) Scarlet (ii) Blue	
1—Forres (1)		Forres (1)	Elgin
2—Elgin (1)		Elgin (1)	—
3—Elgin (1)		Elgin (1)	—
4—Rothes (1)		Rothes (1)	—
5—	—	Fochabers (1)	—
6—	—	Carr Bridge (1)	—
7—	—	(Duff) Urquhart (1)	—
8—	—	Garmouth (1)	—
9—	—	Grantown (1)	—

Became 3 VB Seaforth Highlanders (193) and 6 (Morayshire) Bn Seaforth Highlanders in 1908.

Essex (12)	1st (i) Green (ii) Green	2nd (i) Green (ii) Green
	3rd (i) Green (ii) Black	4th (i) Green (ii) Green
1—Romford	Romford (3)	Braintree
2—Ilford	Ilford (3)	Silvertown
3—Brentwood	Brentwood (3)	Brentwood
4—Chelmsford	Chelmsford (1)	Plaistow
5—Plaistow	Plaistow	—
6—Colchester	Colchester (1)	—
7—Rochford	Rochford (3)	—
8—Stratford	—	—
9—Silvertown	Silvertown	—
10—Witham	Witham (1)	—
11—Dunmow	—	—
12—Braintree	Braintree (1)	—
13—Dedham	—	—
14—Manningtree	—	—
15—Hornchurch	Hornchurch (3)	—
16—Great Bentley	Great Bentley (1)	—
17—Saffron Walden	Saffron Walden†	—
18—Chipping Ongar	Chipping Ongar (3)	—
19—Epping	Epping (3)	—
20—Haverhill	—	—
21—Brentwood	Brentwood (3)*‡	—
22—Waltham Abbey	Waltham Abbey	—
23—Maldon (1)	Maldon (1)	—
24—	Woodford (3)	—

† *United with 2 A Bn Cambridgeshire RV by 1872.*
‡ *United with 2 A Bn Hertfordshire RV by 1872.*

Became 1 VB Essex Regt (79) and 4 Bn Essex Regt in 1908.
2 VB Essex Regt (80) and 5 Bn Essex Regt in 1908.

RIFLE VOLUNTEER CORPS

3 VB Essex Regt (81) and 6 Bn Essex Regt in 1908.
4 VB Essex Regt (82) and 7 Bn Essex Regt in 1908.
New unit recruited from Essex in 1908—Essex and Suffolk Cyclist Bn.

1861	1872	1880
Fifeshire (79)	(i) Scarlet (ii) Blue	
1—Dunfermline (1)	Dunfermline (1)	St Andrews
2—Cupar (1)	Cupar (1)	—
3—E Anstruther (1)	E Anstruther (1)	—
4—Colinsburgh (1)	Colinsburgh (1)	—
5—St Andrews (1)	St Andrews (1)	—
6—Stratheven (1)	Leslie (1)	—
7—Kirkcaldy (1)	Kirkcaldy (1)	—
8—Auchterderran (1)	Lochgelly (1)	—
9—Newburgh	Newburgh (1)	—

By 1872 1st Kinross Corps attached to 1st Fifeshire Administrative Bn.
1 Admin Bn, Fifeshire RV became 6 VB (Fifeshire) The Black Watch (211) in 1887 and 7 (Fife) Bn Black Watch in 1908.
Recruits were also taken from Fifeshire for the Highland Cyclists Bn.

Flintshire (86)	(i) Scarlet (ii) Green	
1—Mold (1)	Mold (1)	Rhyl
2—Hawarden (1)	Hawarden (1)	—
3—Rhyl (1)	Rhyl (1)	—
4—Holywell (1)	Holywell (1)	—
5— —	Flint (1)	—

Became 2 VB R Welsh Fusiliers (216) and 5 (Flintshire) Bn R Welsh Fusiliers in 1908.

Forfar (46)	1st (i) Scarlet (ii) Blue 3rd (i) Scarlet (ii) Blue	2nd (i) Scarlet (ii) Blue
1—Dundee	(Dundee)	Dundee
2—Forfar	Forfar (2)	Forfarshire, or Angus RV
3—Arbroath	Arbroath (1)	Dundee Highland
4— —	—	—
5—Montrose	Montrose (1)	—
6— —	—	—
7—Brechin	Brechin (1)	—
8—Newtyle	(Wharncliffe) (2)	—
9—Glamis	Glamis (2)	—
10—Dundee	(Dundee Highland)	—
11—Taunadin	—	—

	1861		1872	1880
12—	Kirriemuir		Kirriemuir (2)	—
13—	—		Friockheim (1)	—
14—	—		—	—
15—	—		Cortachy (2)	—

Became 2 (170) and 3 (171) VB The Black Watch and 5 (Angus and Dundee) Bn The Black Watch in 1908.

Glamorgan (27) 1st (i) Scarlet (ii) Blue 2nd (i) Scarlet (ii) D Blue
3rd (i) Scarlet (ii) Green

1—Margam	Margam (1)	Margam, Taibach
2—Dowlais	Dowlais (2)	Cardiff
3—Swansea	Swansea	Swansea
4—Swansea	Swansea	—
5—Penllergare	Penllergare†	—
6—Swansea	Swansea (1)	—
7—Taibach	Taibach (1)	—
8—Aberdare	Aberdare (2)	—
9—Baglan	Cwm Avon (1)	—
10—Cardiff	Cardiff (2)	—
11—Bridgend	Bridgend (1)	—
12—Merthyr Tydvil	Merthyr Tydvil (2)	—
13—Llandaff	Llandaff (2)	—
14—Aberdare	Aberdare (2)	—
15—Neath	Neath (1)	—
16—Cardiff	(Bute) Cardiff (2)	—
17—Cadoxton	Cadoxton (1)	—
18—Cowbridge	Cowbridge (1)	—
19— —	Newbridge (2)	—

† *Attached to No 3 Corps.*
3 VB Welsh Regt (125) became 5 Bn Welsh Regt in 1908.
3 Glamorgan VRC (126) became 6 (Glamorgan) Bn Welsh Regt in 1908. Recruits from Glamorgan also joined 7 (Cyclists) Bn Welsh Regt from 1908.
2 VB Welsh Regt (124) became RHA and RA in 1908.

Gloucestershire (2) 1st (i) Green (ii) Green
2nd (i) Green (ii) Scarlet

1—(City of Bristol)	(City of Bristol)	(City of Bristol)
2—Gloucester Dock (2)	Gloucester Dock (1)	Gloucester
3—Gloucester (2)	Gloucester (1)	—
4—Stroud (2)*	—	—
5—Stroud (2)	Stroud (1)	—
6—Stroud (2)	—	—

RIFLE VOLUNTEER CORPS

1861	1872	1880
7—Cheltenham (3)	—	—
8—Tewkesbury (2)	Tewkesbury (1)	—
9—Cirencester (2)	Cirencester (1)	—
10—Cheltenham (3)	Cheltenham (1)	—
11—Dursley (2)	Dursley (1)	—
12—Forest of Dean (2)	Forest of Dean (1)	—
13—Cheltenham (3)	Cheltenham (1)	—
14—Cheltenham (3)	—	—
15—Stow-on-the-Wold	Stow on-the-Wold (1)	—
16—Moreton-in-the-Marsh	Campden (1)	—

Became 1 VB Gloucestershire Regt (100) and 4 (City of Bristol) Bn Gloucestershire Regt in 1908.
2 VB Gloucestershire Regt (101) and 5 Bn Gloucestershire Regt in 1908.
3 VB Gloucestershire Regt (102) and 6 Bn Gloucestershire Regt in 1908.

Haddington (57)	(i) Green (ii) Scarlet	
1—Haddington (1)	Haddington (1)	Haddington
2—Gifford (1)	Gifford (1)	—
3—Haddington (1)	Haddington (1)	—
4—Aberlady (1)	Aberlady (1)	—
5—East Linton (1)	East Linton (1)	—
6— —	Dunglass (1)	—
7— —	North Berwick (1)	—

Became 6 (140) and 7 (188) VB R Scots and 8 Bn R Scots in 1908.

Hampshire (44) 1st (i) Scarlet (ii) Black   2nd (i) Grey (ii) Green
                3rd (i) Scarlet (ii) Yellow

1—Winchester (1)	Winchester (1)	Winchester
2—Southampton	Southampton (4)	Southampton
3—Lymington	Lymington (4)	Portsmouth
4—Havant (2)	Havant (2)	—
5—Portsmouth (2)	(The Portsmouth RVC) (2)	—
6—Gosport (2)	Gosport (2)	—
7—Fareham (3)	Fareham (2)	—
8—Bitterne (3)	Botley (1)	—
9— —	—	—
10—Christchurch	Christchurch (4)	—
11—Romsey (1)	Romsey (1)	—
12—Petersfield (3)	Petersfield (2)	—
13—Andover (1)	Andover (1)	—
14—Lyndhurst	Lyndhurst (4)	—

1861	1872	1880
15—Yateley (1)	Hartley-Wintney (1)	—
16—Alresford (1)	Alresford (1)	—
17—Titchfield (3)	Titchfield (2)	—
18—Basingstoke (1)	—	—
19—Bournemouth	Bournemouth (4)	—
20—Wickham (3)	Wickham (2)	—
21—Alton	Alton (1)	—
22—Bishop's Waltham (3)	—	—
23—Cosham (2)	Portchester (2)	—

Became 1 VB Hampshire Regt (162) and 4 Bn Hampshire Regt in 1908.
2 VB Hampshire Regt (163) and 5 Bn Hampshire Regt in 1908.
3 (DCO) VB Hampshire Regt (164) and 6 Bn (DC) Hampshire Regt in 1908.
4 VB Hampshire Regt (165) and 7 Bn Hampshire Regt in 1908.

Haverfordwest (56)
1—Haverfordwest † —

† United with 1 Admin Bn Pembrokeshire RV by 1872.

Herefordshire (6)	(i) Scarlet (ii) Black	
1—Hereford	Hereford (1)	Hereford
2—Ross	Ross (1)	—
3—Ledbury	Ledbury (1)	—
4—Bromyard	Bromyard (1)	—
5—South Archenfield	South Archenfield (1)	—
6—Leominster	Leominster (1)	—
7—Kington	Kington (1)	—
8—Hereford	Hereford (1)	—

1 Hereford RVC (205) became 1 Bn Herefordshire Regt in 1908.

1861	1872	1880
Hertfordshire (51)	1st (i) Grey (ii) Scarlet	
	2nd (i) Grey (ii) Green	
1—Hertford (2)	Hertford (2)	(Herts and Essex)
2—Watford (1)	Watford (1)	Little Gaddesden, Gt Berkhampstead
3—St Albans (1)	St Albans (1)	—
4—Ashridge (1)	Ashridge (1)	—
5—Hemel Hempstead (1)	Hemel Hempstead (1)	—
6—Bishop's Stortford	Bishop's Stortford (2)	—
7—Berkhampstead (1)	Berkhampstead (1)	—
8—Tring (1)	—	—
9—Ware (2)	Ware (2)	—
10—Royston (2)	Royston (2)	—
11—Cheshunt (2)	—	—
12—Hitchin	—	—

22 Essex RVC was united with 2 Admin Bn Hertfordshire RVC by 1872. Became 1 (181) and 2 (182) VB Bedfordshire Regt, and 1 Bn Hertfordshire Regt in 1908.

Huntingdon (71)
1—Huntingdon †   —
† *United with 1 Huntingdon Light Horse by 1872.*

Recruiting from Huntingdon after 1908 was into 5 Bn Bedfordshire Regt.

Inverness-shire (4)	(i) Scarlet (ii) Buff	
1—Inverness (1)	Inverness (1)	(Inverness Highland)
2—Fort William (1)	Fort William (1)	—
3—Inverness (1)	Inverness (1)	—
4—Inverness (1)	Inverness (1)	—
5—Inverness (1)	Inverness (1)	—
6— —	Kingussie (1)	—
7— —	Beauly (1)	—
8— —	Portree (1)	—
9— —	Campbelltown (1)	—
10— —	Roy Bridge (1)	—

Became 1 VB Cameron Highlanders (155), and 4 Bn Cameron Highlanders in 1908.

RIFLE VOLUNTEER CORPS

1861	1872	1880
Isle of Man (92)	(i) Scarlet (ii) Blue	
1—Castletown	—	—
2—Douglas	Douglas	Douglas†
3—Ramsey	—	—

† *Attached to 15 Lancashire RVC by 1880.*

Isle of Wight (58)	(i) Green (ii) Light Green	
1—Ryde (1)	Ryde (1)	Newport
2—Newport (1)	Newport (1)	—
3—Ryde (1)	—	—
4—Nunwell (1)	Nunwell (1)	—
5—Ventnor (1)	Ventnor (1)	—
6—Sandown (1)	—	—
7—Cowes (1)	Cowes (1)	—
8—Freshwater (1)	—	—

Became 5 VB Hampshire Regt (189), and 8 (Isle of Wight 'Princess Beatrice's') Bn Hampshire Regt in 1908.

Kent (26) 1st (i) Green (ii) Green 2nd (i) Green (ii) Scarlet
3rd (i) Green (ii) Black 4th (i) Green (ii) Scarlet
5th (i) Green (ii) Green

1—Maidstone (3)	Maidstone (3)	Tunbridge
2— —	—	(East Kent) Canterbury
3—Lee (1)	Lee (1)	Blackheath
4—Woolwich (1)	Woolwich†	Woolwich Arsenal
5—Canterbury (4)	Canterbury (4)	(Weald of Kent)
6—Canterbury (4)	Canterbury (4)	—
7—Kidbrook (1)	—	—
8—Sydenham (1)	—	—
9—Chatham (3)	Chatham (3)	—
10— —	—	—
11—Farnborough (2)	—	—
12—Dartford (3)	Dartford (3)	—
13—Greenwich (1)	Greenwich (1)	—
14—Tunbridge (2)	Tunbridge (2)	—
15—Sutton (3)	Sutton Valence (3)	—
16—Sittingbourne (4)	Sittingbourne (4)	—
17—Tunbridge Wells (2)	Tunbridge Wells (2)	—
18—Bromley (1)	Bromley (1)	—
19—Rochester (3)	Rochester (3)	—
20—Northfleet (3)	—	—
21—Lewisham (1)	—	—

RIFLE VOLUNTEER CORPS

1861	1872	1880
22—Sheerness (3)	—	—
23—Penshurst (2)	Penshurst (2)	—
24—Ash (4)	—	—
25—Blackheath (1)	Blackheath (1)	—
26—(Royal Arsenal)	(Royal Arsenal)	—
27—Deptford (1)	Deptford (1)	—
28—Charlton (1)	Charlton (1)	—
29—Ashford (4)	Ashford (4)	—
30—	—	—
31—Leeds Castle (3)	Leeds Castle (3)	—
32—Eltham (1)	Eltham (1)	—
33—Sevenoaks (2)	Sevenoaks (2)	—
34—Deptford (1)	Deptford (1)	—
35—Westerham (2)	Westerham (2)	—
36—Wingham (4)	Wingham (4)	—
37—Cranbrook	Cranbrook (5)	—
38—	Hawkhurst (5)	—
39—Malling (2)	West Malling (3)	—
40—	Staplehurst (5)	—
41—	Goudhurst (5)	—
42—	Brenchley (5)	—
43—	Rolvedon (5)	—
44—	—	—
45—	Rochester (3)	—

† *Attached to 26 (Royal Arsenal) Kent RVC by 1872.*

Became 1 (119) and 2 (122) VB The Buffs, and 4 Bn East Kent Regt (The Buffs) and 5 (The Weald of Kent) Bn East Kent Regt in 1908.

1 VB R West Kent Regt (118) and 4 (The Queen's Own) Bn R West Kent Regt and 5 Bn R West Kent Regt in 1908.

Recruits from Kent also joined Kent Cyclist Bn.

4 VB R West Kent Regt (123) was disbanded by 1908.

Kincardineshire (55)	(i) Green (ii) Green	
1—Feteresso	—	(Deeside Highland RV) Banchory
2—Banchory	Banchory (1)	—
3—Lawrencekirk	Laurencekirk (1)	—
4—Fettercairn	—	—
5—Auchinblay	Auchinblae (1)	—
6—Netherley	Portlethen (1)	—
7—	Durris (1)	—
8—	Maryculter (1)	—

Became 5 (Deeside Highland) VB Gordon Highlanders (187) and 7 (Deeside Highland) Bn Gordon Highlanders.

	1861	1872	1880
Kinross-shire (93)			
1—Kinross		—	—

United with 1st Admin Bn Fifeshire RV by 1872.

Kirkudbrightshire (77)
	1861	1872	1880
1—	Kirkudbright (1)	Kirkudbright (1)	(See Galloway)
2—	Castle Douglas (1)	Castle Douglas (1)	—
3—	Galloway (1)	New Galloway (1)	—
4—	Gatehouse (1)	—	—
5—	Maxwelton (1)	Maxwelton (1)	—
6—	—	Dalbeattie (1)	—

After 1908, recruiting was into 5 Bn KOSB.

Lanarkshire (25) 1st (i) Scarlet (ii) Blue   3rd (i) Scarlet (ii) Blue
                  4th (i) Scarlet (ii) Green   5th (i) Scarlet (ii) Buff
                  6th (i) Scarlet (ii) Yellow 7th (i) Scarlet (ii) Yellow
                  8th (i) Scarlet (ii) Blue   9th (i) Scarlet (ii) Blue
                 10th (i) Scarlet (ii) Blue

	1861	1872	1880
1—	Glasgow	Glasgow	Glasgow
2—	—	—	—
3—	Glasgow	Glasgow	Glasgow
4—	(Glasgow 1st Northern)	(Glasgow 1st Northern)	(Glasgow 1st Northern)
5—	Glasgow	Glasgow	(Glasgow 2nd Northern)
6—	—	—	Overnewtown
7—	—	—	Airdrie
8—	—	—	(The Blythswood)
9—	—	—	Lanark

	1861	1872	1880
10—	—	—	(Glasgow Highland)
11—	—	—	—
12—	—	—	—
13—	—	—	—
14—	—	—	—
15—	—	—	—
16—	Hamilton (3)	Hamilton (1)	—
17—	—	—	—
18—	—	—	—
19—	(Glasgow 2nd Northern)	(Glasgow 2nd Northern)	—
20—	—	—	—
21—	—	—	—
22—	—	—	—
23—	—	—	—
24—	—	—	—
25—	Glasgow (6)	Glasgow	—
26—	Glasgow (6)	—	—
27—	Glasgow (6)	—	—
28—	—	—	—
29—	Coatbridge	Coatbridge	—
30—	Glasgow (4)	—	—
31—	Glasgow (4)	(The Blythswood)	—
32—	Summerlee	Summerlee (4)	—
33—	—	—	—
34—	—	—	—
35—	—	—	—
36—	—	—	—
37—	Lesmahagow (8)	Lesmahagow (3)	—
38—	Glasgow (4)	—	—
39—	—	—	—
40—	Glasgow (6)	—	—
41—	—	—	—
42—	Uddingstone (3)	—	—
43—	Gartsherrie	—	—
44—	Blantyre (3)	—	—
45—	Glasgow (4)	—	—
46—	Glasgow (4)	—	—
47—	Glasgow (4)	—	—
48—	Airdrie	—	—
49—	Lambhill	—	—
50—	—	—	—
51—	—	—	—
52—	Hamilton (3)	—	—
53—	—	—	—
54—	—	—	—

1861	1872	1880
55—Lanark (8)	Lanark (3)	—
56—Bothwell (3)	Bothwell (1)	—
57—Wishaw (3)	Wishaw (1)	—
58— —	—	—
59— —	—	—
60—Glasgow	—	—
61—Glasgow	—	—
62— —	Biggar (3)	—
63— —	—	—
64— —	—	—
65— —	—	—
66— —	—	—
67— —	—	—
68—Glasgow (6)	—	—
69—Glasgow (6)	—	—
70—Glasgow (6)	—	—
71—Glasgow (6)	—	—
72— —	—	—
73—Carluke (8)	Carluke (3)	—
74— —	—	—
75—Glasgow (4)	—	—
76— —	—	—
77— —	—	—
78—Glasgow	—	—
79— —	—	—
80— —	—	—
81— —	—	—
82— —	—	—
83— —	—	—
84—Glasgow (4)	—	—
85— —	—	—
86—Glasgow	—	—
87— —	—	—
88—Glasgow	—	—
89— —	—	—
90— —	—	—
91— —	—	—
92— —	—	—
93—Glasgow	—	—
94—Douglas (8)	Douglas (3)	—
95—Bailleston*	Bailliestown (4)	—
96—Glasgow	—	—
97— —	Woodhead (4)	—
98— —	Wattston (4)	—
99— —	Calderoruix (4)	—
100— —	Calderoruix (4)	—

RIFLE VOLUNTEER CORPS

	1861	—	1872	1880
101—	—		Newarthill (4)	—
102—	—		Motherwell (1)	—
103—	—		East Kilbride (1)	—
104—	—		Mossend, Holytown, Bellshill (4)	—
105—	—		(Glasgow Highland)	—

Became 2 VB Scottish Rifles (110) and 6 Bn Scottish Rifles in 1908.
9 Lanarkshire VRC (116) became 8 (Lanark) Bn HLI in 1908.
1 Lanarkshire VRC (109) became 5 Bn Scottish Rifles in 1908.
3 Lanarkshire VRC (111) became 7 Bn Scottish Rifles in 1908.
4 Lanarkshire VRC became 4 VB Scottish Rifles (112) and 8 Bn Scottish Rifles in 1908.
1 VB HLI (113) became 5 (City of Glasgow) Bn HLI in 1908.
2 VB HLI (114) became 6 (City of Glasgow) Bn HLI in 1908.
3 (Blythswood) VB HLI (115) became 7 (Blythswood) Bn HLI in 1908.
5 (Glasgow Highland) VB HLI (117) became 9 (Glasgow Highland) Bn HLI in 1908.

Lancashire (3)
1st (i) Green   (ii) Black   2nd (i) Scarlet   (ii) White
4th (i) Scarlet (ii) Black   5th (i) Scarlet   (ii) Yellow
8th (i) Grey   (ii) Black   9th (i) Scarlet   (ii) Green
10th (i) Scarlet (ii) Blue   11th (i) Scarlet (ii) White
13th (i) Scarlet (ii) Blue   15th (i) Scarlet (ii) Blue
21st (i) Green (ii) Scarlet   23rd (i) Scarlet (ii) Green
24th (i) Scarlet (ii) Blue   27th (i) Scarlet (ii) Green
33rd (i) Green (ii) Scarlet   40th (i) Grey   (ii) Scarlet
47th (i) Green (ii) Scarlet   48th (i) Green   (ii) Black
56th (i) Scarlet (ii) Blue   64th (i) Green   (ii) Scarlet
80th (i) Scarlet (ii) Blue

1—Liverpool (1)	Liverpool	Liverpool
2—Blackburn	Blackburn (8)	Blackburn
3—	—	—
4—Rossendale (3)	Rossendale (3)	Burnley
5—Liverpool (2)	(Liverpool RV Bde)	(Liverpool RV Bde)
6—(1st Manchester)	(1st Manchester)	—
7—Accrington (3)	Accrington (3)	—
8—Bury	Bury	Bury
9—Warrington	Warrington (9)	Warrington
10—Lancaster	Lancaster (5)	Ulverston
11—Preston	Preston (6)	Preston
12—	—	—
13—Southport	Attached to 1 Corps	Southport
14—Edge Hill (2)	—	—
15—Liverpool	Liverpool	Liverpool†

1861	1872	1880
16— —	—	—
17—Burnley (3)	Burnley (3)	—
18— —	—	—
19—Liverpool (2)	—	—
20— —	—	—
21—Wigan	Wigan (4)	Manchester
22—Liverpool (1)	—	—
23—Ashton-under-Lyne	Ashton (7)	—
24—Rochdale	Rochdale	Rochdale
25—Liverpool	—	—
26—Haigh	—	—
27—Bolton	Bolton	Bolton
28—(2nd Manchester)	—	—
29—Lytham	Lytham (3)	—
30— —	—	—
31—Oldham	Oldham (7)	—
32—Liverpool	—	—
33—Ardwick	(2nd Manchester)	(2nd Manchester)
34— —	—	—
35— —	—	—
36—Accrington (3)	—	—
37—(North Lonsdale)	37a-Ulverston (5)	—
38—Fairfield (1)	37b-Barrow (5)	—
39—Liverpool (2)	37c-Hawkshead (5)	—
40—(3rd Manchester)	(3rd Manchester)	(3rd Manchester)
41—Liverpool	—	—
42—Childwall	—	—
43—Fallowfield	—	—
44—Longton	—	—
45—Liverpool (1)	—	—
46—Swinton	Swinton (4)	—
47—St Helens	St Helens	St Helens
48—Prescot	Attached to 1 Corps	Attached to 1 RVC
49—Newton-le-Willows	Newton-le-Willows (9)	—
50— —	—	—
51—Liverpool	—	—
52—Dalton*	—	—
53— —	Cartmel (5)	—
54—Ormskirk	Attached to 1 Corps	—
55—Leigh	Leigh (4)	—
56—Salford	Salford	Salford
57—Ramsbottom	Ramsbottom (3)	—
58— —	—	—
59—Leyland	Leyland (6)	—
60—Atherton	Atherton (4)	—
61—Chorley	—	—

Page 107   *Bandsman's pouch of the 2 VB Suffolk Regt, post 1881. Glengarry badges of 2 VB Suffolk Regt and 1 VB R Berkshire Regt*

Page 108 *Cap badges: Loyal Suffolk Hussars, Duke of Lancaster's Own, London Rifles Brigade 1920–56*

RIFLE VOLUNTEER CORPS

1861	1872	1880
62—Clitheroe	Clitheroe (8)	—
63—Toxteth Park (2)	—	—
64—Liverpool (2)	(Liverpool Irish)	(Liverpool Irish)
65—Rossall	Rossall (5)	—
66—Liverpool (1)	—	—
67—Worsley	Worsley (4)	—
68—Liverpool (2)	—	—
69—Liverpool (1)	—	—
70—Droylesden	—	—
71—Liverpool (2)	—	—
72—Old Swan	—	—
73—Newton	—	—
74—Liverpool	—	—
75—	—	—
76—Farnworth	Farnworth (4)	—
77—Widnes	—	—
78—(4th Manchester)	—	—
79—Liverpool*	—	—
80—Liverpool	(Liverpool Press Guard)	(Liverpool Press Guard)
81—	Wheetton (8)	—
82—	Hindley—Attached to 27 Corps	—
83—	Knowsley—Attached to 1 Corps	—
84—	Padiham (3)	—
85—	—	—
86—	—	—
87—	—	—
88—	Haslingden (3)	—

† *The 13 Lancs RVC and 2 Isle of Man RVC were attached to 15 Corps by 1880.*

37 North Lonsdale Lancs RVC became 1 VB R Lancs (39) in 1883 and 4 Bn (The KO) R Lancs Regt in 1908.

2 VB R Lancs Regt (52) became 5 Bn (The KO) R Lancs Regt in 1908.

1 VB K Liverpool Regt (30) became 5 Bn (The Kings) Liverpool Regt in 1908.

4 VB K Liverpool Regt (44) became 7 Bn (The Kings) Liverpool Regt in 1908.

5 (Irish) VB K Liverpool Regt (47) became 8 (Irish) Bn (The Kings) Liverpool Regt in 1908.

19 Lancs RVC, 80 Lancs (Liverpool Press Guard) became 6 VB K Liverpool Regt (48) and 9 Bn (The Kings) Liverpool Regt in 1908.

8 (Scottish) VB K Liverpool Regt (53) became 10 (Scottish) Bn (The

Kings) Liverpool Regt in 1908.

1 Isle of Man RV renamed 7 (Isle of Man) VB Liverpool Regt after 1908 (219).

2 VB K Liverpool Regt (34) became 6 Bn (The Kings) Liverpool Regt in 1908.

3 VB K Liverpool Regt (42) was disbanded prior to 1908.

9 Lancs RVC became 1 VB Lancs Regt (38) in 1886 and 4 Bn S Lancs Regt in 1908.

2 VB S Lancs Regt (5) became 5 Bn S Lancs Regt in 1908.

1 VB Loyal N Lancs (40) became 4 Bn Loyal N Lancs in 1908.

27 Lancs RVC became 14 Lancs RVC in 1881, 2 (43) VB Loyal N Lancs in 1883 and 5 Bn Loyal N Lancs in 1908.

1 VB Lancs Fusiliers (37) became 5 Bn Lancs Fusiliers in 1908.

2 VB Lancs Fusiliers (41) became 6 Bn Lancs Fusiliers in 1908.

3 VB Lancs Fusiliers (46) became 7 and 8 Bn Lancs Fusiliers in 1908.

2 VB E Lancs (31) became 4 Bn E Lancs in 1908.

3 VB E Lancs (32) became 5 Bn E Lancs in 1908.

4 Lancs RVC became 1 VB Manchester (33) and 5 Bn Manchester Regt in 1908.

2 VB Manchester (35) became 6 Bn Manchester Regt in 1908.

4 VB Manchester (45) became 7 Bn Manchester Regt in 1908.

5 VB Manchester (49) became 8 (Ardwick) Bn Manchester Regt in 1908.

3 VB Manchester (36) became 9 Bn Manchester Regt in 1908.

6 VB Manchester (51) became 10 Bn Manchester Regt in 1908.

1861	1872	1880
Leicestershire (31)	(i) Scarlet (ii) White	
1—Leicester (1)	Leicester (1)	Leicester
2—Belvoir (1)	Belvoir (1)	—
3—Melton Mowbray (1)	Melton Mowbray (1)	—
4—Leicester (1)	Leicester (1)	—

1861	1872	1880
5—Leicester (1)	Leicester (1)	—
6—Loughborough (1)	Loughborough (1)	—
7—Lutterworth (1)	Lutterworth (1)	—
8—Ashby-de-la-Zouch (1)	Ashby-de-la-Zouch (1)	—
9—Leicester (1)	Leicester (1)	—
10—Hinckley (1)	Hinckley (1)	—

Became 1 VB Leicestershire Regt (138) and 4 and 5 Bns Leicestershire Regt in 1908.

Lincolnshire (42)	1st (i) Scarlet (ii) Blue	
	2nd (i) Scarlet (ii) Blue	
1—Lincoln (1)	Lincoln (1)	Lincoln
2—Louth (1)	Louth (1)	Grantham
3—Grantham (2)	Grantham (2)	—
4—Boston (3)	Boston (2)	—
5—Stamford (2)	Stamford (2)	—
6—Great Grimsby (1)	Great Grimsby (1)	—
7—Spilsby (1)	Spilsby (1)	—
8—Sleaford (2)	Sleaford (2)	—
9—Horncastle (1)	Horncastle (1)	—
10—	—	—
11—Alford (1)	Alford (1)	—
12—Barton-upon-Humber (1)	Barton-upon-Humber (1)	—
13—Spalding (3)	Spalding (2)	—
14—Swineshead (3)	—	—
15—Bourne (2)	Bourne (2)	—
16—Holbeach (3)	—	—
17—Donington (3)	Donington (2)	—
18—Folkington (2)	Folkington (2)	—
19—Gainsborough (1)	Gainsborough (1)	—
20—Market Raisen (1)	Market Raisen (1)	—

Became 1 (158), 2 (159) and 3 (160) VB's Lincolnshire Regt, and 4 and 5 Bns Lincolnshire Regt in 1908.

Linlithgowshire (82)	(i) Green (ii) Scarlet	
1—Linlithgow	Linlithgow (1)	Linlithgow
2—Boness	Boness (1)	—
3—Bathgate	Torphichen (1)	—
4— —	Bathgate (1)	—
5— —	Uphall (1)	—

Became 8 VB R Scots (214) and 10 (Cyclist) Bn R Scots in 1908.

	1861	1872	1880
London (49)	1st (i) Green	(ii) Green	2nd (i) Green (ii) Scarlet
	3rd (i) Scarlet	(ii) Buff	
1—	(City of London)	(City of London)	(City of London)
2—	Little New Street	Fleet Street	Holborn Circus
3—	—	Farringdon Street	Farringdon Street

Under 2 and 3 Corps, the drill hall addresses are given.

1 London VRC (176) became 5 (City of London) Bn London Regt in 1908, attached to Rifle Bde in 1937.

2 London VRC (177) became 6 (City of London) Bn London Regt in 1908, converted to RE in 1937.

3 London VRC (178) became 7 (City of London) Bn London Regt in 1908.

Merionethshire (29)

1—Bala	Dolgelly (United with 1 A Bn Montgomery RV by 1872)	—
2—Dolgelly	—	—
3—Corwen	—	—

Middlesex (2)

	1861	1872	1880
1—	(Victoria)	(Victoria)	(Victoria) (Att to 11) (i) Green (ii) Black
2—	(South Middlesex)	(South Middlesex)	(South Middlesex) (i) Grey (ii) Scarlet
3—	Hampstead (2)	Hampstead (2)	Hornsey (i) Grey (ii) Scarlet
4—	Islington (1)	(West London)	(West London) (i) Grey (ii) Scarlet
5—	—	—	—
6—	—	—	—
7—	Islington (1)	—	—
8—	—	—	—
9—	(West Middlesex)	(West Middlesex)	(West Middlesex) (i) Grey (ii) Scarlet
10—	—	—	—
11—	(St George's)	(St George's)	(St George's) (i) Green (ii) Black
12—	Barnet (6)	Barnet (2)	—
13—	Hornsey (2)	Hornsey (2)	—
14—	Highgate (2)	Highgate (2)	—
15—	(London Scottish)	(London Scottish)	(London Scottish) (i) Grey (ii) Blue
16—	Hounslow	Hounslow (7)	(SW Middlesex) (i) Grey (ii) Grey

1861	1872	1880
17— —	—	—
18—Harrow	Harrow (Att to 9)	Harrow (Att to 9) (i) Green (ii) Green
19—Bloomsbury	Bloomsbury	Bloomsbury (i) Scarlet (ii) D Blue
20—Euston Square (4)	Euston Square	Euston Square (i) Grey (ii) Scarlet
21—(Civil Service)	(Civil Service)	(Civil Service) (i) Grey (ii) Blue
22—(Queen's)	(Queen's)	(Queen's) (i) Grey (ii) Blue
23—(Inns of Court)	(Inns of Court)	(Inns of Court) (i) Grey (ii) Scarlet
24—Uxbridge	Uxbridge (7)	—
25— —	—	—
26—Custom House (5)	(Customs and Docks)	(Customs and Docks) (i) Green (ii) Scarlet
27— —	—	—
28—(London Irish)	(London Irish)	(London Irish) (i) Green (ii) Lt Green
29—Regent's Park	(North Middlesex)	(North Middlesex) (i) Green (ii) Black
30—Ealing	Ealing (7)	—
31— —	—	—
32—Conduit Street	—	—
33—Tottenham (6)	Tottenham (2)	—
34— —	—	—
35—Enfield (6)	—	—
36—Paddington	Paddington	Paddington (i) Green (ii) Black
37—Local Board of Works (4)	(St Giles's and St George's)	(St Giles's and St George's) (i) Green (ii) Green
38—Burlington House	The Arts Club	(Artists') (i) Grey (ii) Grey
39—Clerkenwell (3)	(The Finsbury RVC)	(The Finsbury RVC) (i) Green (ii) Scarlet
40—Gray's Inn (3)	(Central London Rifle Rangers)	(Central London Rangers) (i) Green (ii) Scarlet
41—Enfield (6)	Enfield (2)	—
42—St Katherine's Docks (5)	—	—
43—Hampton	Sunbury (7)	—
44—Staines	Staines (7)	—

	1861	1872	1880
45—	Sunbury	—	—
46—	—	Westminster	Westminster (i) Scarlet (ii) Blue
47—	—	—	—
48—	—	Lincoln's Inn	—
49—	—	Post Office (formed 1868)	Post Office (i) Grey (ii) Blue
50—	—	—	(Bank of England) (i) (ii) Rifle Green

The Middlesex Rifle Volunteer Corps were re-numbered again after 1881. They were attached to the Middlesex Regt, R Fusiliers, KRRC and Rifle Brigade as Volunteer battalions during the Territorial re-organisation of the numbered regular regiments, but many corps retained their old titles, and continued to hold them after the next re-organisation when the Territorial Force was formed in 1908.

The 1881 Middlesex Corps are given below, the number in brackets being the pre-1881 distinction, followed by each corps' subsequent history.

1 Victoria and St George's (1, 11). Attached to KRRC in 1881 as 6 Middlesex RV, became 1 Middlesex RV in 1892, amalgamated with 19 Corps in 1908 to become 9 Bn The London Regt (Queen Victoria's Rifles).
2 South Middlesex (2). Attached to KRRC in 1881.
3 Hornsey (3). Became 7 Bn The London Regt in 1908, amalgamated with 8 Bn (see 24 Corps) in 1922 to become 7 Bn (Post Office Rifles), converted to RE in 1937.
4 West London (4). 7 Middlesex RV was merged in 4 in 1861, it was attached to KRRC as 4 Middlesex (Kensington Rifles) in 1908, after having been re-titled Kensington Rifles in 1905. It was called Kensington Regt in 1909, Princess Louise's title was granted in 1914 and attached to the Middlesex Regt in 1937.
5 West Middlesex (9). 9 Corps was attached to 5 Corps and both attached to KRRC in 1881, becoming 9 Bn Middlesex Regt in 1908.
6 St George's (11). (See 1.)
7 London Scottish (15). Attached to Rifle Bde in 1881, became 14 Bn The London Regt (London Scottish) in 1908 and was attached to the Gordon Highlanders in 1937.
8 Middlesex. Became 2 VB Middlesex Regt in 1881.
9 West Middlesex (9). Attached to 5 Middlesex RVC as above.
10 Bloomsbury (19). Attached to KRRC in 1881, became 1 VB R Fusiliers in 1883, 1 Bn London Regt (R Fusiliers) in 1908 and 8 Bn R Fusiliers in 1937.
11 Railways (20). Attached to Middlesex Regt in 1881, became 3 VB R Fusiliers in 1890, 3 Bn The London Regt (R Fusiliers) in 1909 and 10 Bn R Fusiliers in 1937.
12 Civil Service Rifles (21). Amalgamated with 27 (Inland Revenue), 31

RIFLE VOLUNTEER CORPS

(Whitehall) and 34 (Admiralty) Middlesex RV in 1860, attached to KRRC as 12 Middlesex in 1881, became 15 Bn The London Regt (Prince of Wales's Own Civil Service Rifles) in 1908 and joined 16 Bn The London Regt as 16 Bn The London Regt (Queen's Westminster and Civil Service Rifles) in 1922.

13 Queen's (22). Attached to KRRC in 1881, became 16 Bn The London Regt (Queen's Westminster Rifles) in 1908, amalgamating with 15 Bn The London Regt in 1922 to become 16 Bn (Queen's Westminster and Civil Service Rifles).

14 Inns of Court (23). Attached to Rifle Bde in 1881, did not become a part of the London Regt.

15 Customs and Docks (26). Joined (42) in 1860, absorbed 9 Tower Hamlets RV in 1864 and (42) in 1866 to be called 'Customs and Docks'. Absorbed 8 Tower Hamlets in 1868 and, as 15 Middlesex, was attached to Rifle Bde. At the same time, 3, 7, 10 and 11 Tower Hamlets RV amalgamated to become Tower Hamlets RV in 1880. In 1908, 2 Tower Hamlets and 15 Middlesex amalgamated to become 17 Bn The London Regt (Poplar and Stepney Rifles). It was renamed Tower Hamlets Rifles in 1926 and was attached to the Rifle Bde in 1937.

16 London Irish (28). Attached to Rifle Bde in 1881 and became 18 Bn The London Regt (London Irish Rifles) in 1908. Attached to R Ulster Rifles in 1937.

17 North Middlesex (29). Attached to Middlesex Regt in 1881 and became 19 Bn The London Regt (St Pancras) in 1908. Converted to R Engineers in 1937.

18 Paddington (36). Attached to Rifle Bde in 1881, became 10 Bn The London Regt (Paddington Rifles) in 1908 and disbanded in 1912. 10 Bn The London Regt (Hackney) was formed in October 1912 to replace it from 7 Bn Essex Regt and it became 5 Bn R Berks Regt in 1925.

19 St Giles's and St George's (37). Attached to Rifle Bde in 1881, for subsequent history see 1 Corps.

20 Artists (38). Attached to Rifle Bde in 1881 and became 28 Bn The London Regt (Artists' Rifles) in 1908.

21 Finsbury (39). Attached to Rifle Bde in 1881 and became 11 Bn The London Regt (Finsbury Rifles) in 1908, became R Artillery in 1936.

22 Central London Rangers (40). Attached to KRRC in 1881, became 12 Bn The London Regt (The Rangers) in 1908.

23 Westminster (46). Attached to R Fusiliers in 1881, became 2 VB R Fusiliers in 1883, 2 Bn The London Regt (R Fusiliers) in 1908 and 9 Bn R Fusiliers in 1937.

24 Post Office Rifles (49). Attached to Rifle Bde in 1881, became 8 Bn The London Regt (Post Office Rifles) in 1908 and was amalgamated with 7 Bn The London Regt (see 3 Corps) in 1922.

25 Bank of England. Amalgamated with 12 Middlesex and attached to KRRC in 1881 (see 12 Corps).

26 Cyclists. Formed in 1888 and attached to KRRC, transferred to Rifle Bde in 1889, disbanded in 1922 and reformed as R Signals unit.

When the London Regt was formed in 1908 from the Volunteer Bns in the Greater London area, the 26 and 27 Regts remained vacant out of the 28 numbers. The 26th was to have been the HAC infantry, and the 27th the Inns of Court Regt, but they both remained independent. The London Regt was disbanded in 1936/7 when the units were converted into R Artillery, R Engineers and R Signals.

A new unit was formed in Ravenscourt Park in 1908, 10 Bn Middlesex Regt.

1861	1872	1880
Midlothian (32)	2nd (i) Scarlet (ii) Black	
1—(Leith)	(Leith)	—
2—Dalkeith	Dalkeith (1)	Penicuick
3—Penicuick	Penicuick (1)	—
4—Corstorphine	—	—
5—	Musselburgh (1)	—

1, 2 and 3 Peeblesshire RVC were attached to 2, 3 and 5 to form Dalkeith Administrative Bn by 1872.

Became 5 VB R Scots (130) and 7 Bn R Scots in 1908.

Provided 2 Coys to 8 Bn R Scots in 1908.

Monmouthshire (36)	1st (i) Green (ii) Black	
	2nd (i) Scarlet (ii) Grass Green	
	3rd (i) Green (ii) Black	
1—Chepstow (1)	Chepstow (1)	Newport
2—Pontypool	Pontypool	Pontypool
3—Newport (1)	Newport (1)	Pontypool
4—Tredegar (1)	Blaenavon (2)	—
5—Pontypool (2)	Pontypool (2)	—
6—Monmouth (2)	Monmouth (2)	—
7—Newport (2)	Newport (2)	—
8—Usk (2)	Usk (2)	—
9—Abergavenny (2)	Abergavenny (2)	—
10—Risca	Risca (1)	—
11—No title	—	—

1 Corps became 2 VB S Wales Borderers (147) in 1880/1 and 1 Bn Monmouthshire Regt in 1908.

2 Corps became 3 VB S Wales Borderers (148) in 1880/1 and 2 Bn Monmouthshire Regt in 1908.

3 Corps became 4 VB S Wales Borderers (149) in 1880/1 and 3 Bn Monmouthshire Regt in 1908.

1861	1872	1880
Montgomeryshire (73)		
1—Newtown	Newtown (1)	—
2—Welshpool	Welshpool (1)	—
3—Welshpool	Welshpool (1)	—
4—Machynlleth	Llanidloes (1)	—

Became 5 VB S Wales Borderers (209) in 1881 and 7 (Merioneth and Montgomery) Bn R Welsh Fus in 1908.

Nairn (9)
1—Nairn

No territorial infantry units were formed from Nairn in 1908.

Newcastle-upon-Tyne (81) (i) Scarlet (ii) Black
1—Newcastle   Newcastle   Newcastle

Amalgamated with Northumberland after 1880.

Norfolk (17) 1st (i) Scarlet (ii) White   2nd (i) Scarlet (ii) White
              3rd (i) Scarlet (ii) White   4th (i) Grey (ii) Grey

1—(City of Norwich)	(City of Norwich)	(City of Norwich)
2—(Great Yarmouth)	(Great Yarmouth)	(Norfolk and Suffolk Great Yarmouth)
3—	—	East Dereham
4—	—	Norwich
5—King's Lynn	King's Lynn (1)	—
6—Aylsham	Aylsham (1)	—
7—Harleston	Harleston (2)	—
8—Diss	Diss (2)	—
9—Loddon	Loddon (2)	—
10—Fakenham	—	—
11—Holkham	Holkham (1)	—
12—Reepham	Reepham (1)	—
13—Cromer	—	—
14—Stalham	Stalham (2)	—
15—East Dereham	East Dereham (1)	—
16—Swaffham	Swaffham (1)	—
17—Snettisham	Heacham (1)	—
18—Blofield	Blofield (2)	—
19— —	Holt (1)	—
20—Attleburgh	Attleburgh (2)	—
21—Wymondham	Wymondham (2)	—
22—Thetford	Thetford (2)	—

	1861	1872	1880
23—	—	Downham Market (1)	—
24—	—	North Walsham (1)	—

3 VB Norfolk Regt (92) became 5 Bn Norfolk Regt in 1908.
1 (90) and 4 (93) VB Norfolk Regt became 4 Bn Norfolk Regt in 1908.
6 (Cyclist) Bn Norfolk Regt was formed in 1908.
2 VB Norfolk Regt (91) became Norfolk and Suffolk Bde Coy ASC in 1908.

Northamptonshire (15)	(i) Grey (ii) Scarlet	
1—Althorpe (1)	Althorpe (1)	Northampton
2—Towcester (1)	Towcester (1)	—
3—	—	—
4—Northampton (1)	Northampton (1)	—
5—Northampton (1)	Northampton (1)	—
6—Peterborough (1)	Peterborough (1)	—
7—Wellingborough (1)	Wellingborough (1)	—
8—Daventry (1)	Daventry (1)	—
9—	Kettering (1)	—

Became part of 1 VB Northants Regt (88) in 1881 and 4 Bn Northants Regt in 1908, extra troops joined Northampton Batt RFA and E Midland Bde Coy ASC.

Northumberland (13)	1st (i) Grey (ii) Scarlet	
	8th (i) Scarlet (ii) Green	
1—Tynemouth	—	Alnwick
2—Hexham (1)	Hexham (1)	
3—Morpeth (1)	Morpeth (1)	—
4—Wooler-in-Glendale (1)	Belford (1)	
5—Alnwick (1)	Alnwick (1)	—
6—Tynedale (1)	Tynedale (1)	—
7—Allendale (1)	Allendale (1)	—
8—	Walker, Newcastle	Walker, Newcastle
9—	—	—
10—	Lowick (1)	—
11—	St John-lee (1)	—

2–7, 10 and 11 corps and 1 Berwick on Tweed RVC had formed 1 (Alnwick) Adm Bn by 1872.

1 VB Northumberland Fus (83) became 4 Bn Northumberland Fus in 1908.

2 VB Northumberland Fus (84) became 5 Bn Northumberland Fus in 1908.

3 VB Northumberland Fus (213) became 6 Bn Northumberland Fus in 1908.

7 Bn Northumberland Fus was formed in 1908.
Recruits were also taken into Northern Cyclist Bn.

	1861	1872	1880
Nottinghamshire (28)		1st (i) Green (ii) Black	
		2nd (i) Scarlet (ii) Lincoln Green	
1—	(Robin Hood)	(Robin Hood)	(Robin Hood)
2—	East Retford	East Retford (1)	East Retford
3—	Newark	Newark (1)	—
4—	Mansfield	Mansfield (1)	—
5—	Thorney Wood Chase	Thorney Wood Chase (1)	—
6—	Collingham	Collingham (1)	—
7—	Worksop	Worksop (1)	—
8—	Southwell	Southwell (1)	—

The Robin Hood Rifles, 1 Notts VRC (27) became 7 (Robin Hood) Bn Notts and Derby Regt (Sherwood Foresters) in 1908.

4 Notts VB Notts and Derby Regt (128) became 8 Bn Notts and Derby Regt (Sherwood Foresters) in 1908.

Orkney (74)
1—Lerwick   United with 1 Adm Bn Sutherland VRC by 1872.

Oxfordshire (7)		1st (i) Scarlet (ii) Dark Blue	
		2nd (i) Scarlet (ii) White	
1—	(University of Oxford)	(Oxford University)	(Oxford University)
2—	City of Oxford (2)	City of Oxford (1)	Oxford
3—	Banbury (2)	Banbury (1)	—
4—	Henley-on-Thames (2)	Henley-on-Thames (1)	—
5—	Woodstock (2)	—	—

1861	1872	1880
6—Deddington (2)	Deddington (1)	—
7—Bicester (2)	—	—
8—Thame	Thame (1)	—
9— —	Woodstock (1)	—

1 VB Oxfordshire LI (65) became Oxford University OTC in 1908.
2 VB Oxfordshire LI (66) became 4 Bn Ox and Bucks LI in 1908.

Peeblesshire (91)
1—Peebles	Peebles	—
2—Broughton	Broughton	—
3—Inverleithen	Inverleithen	—
4—Linton	—	—

All three corps united with 1 Adm Bn Midlothian RV by 1872.

Pembrokeshire (5) (i) Scarlet (ii) Dark Blue
1—Milford	Milford (1)	Haverfordwest
2—Pembroke Dock	—	—
3—Pembroke	Pembroke (1)	—

By 1872, the 1st Adm Bn consisted of the two Pembroke Corps, the 2nd Cardigan and 1st Haverfordwest RVCs.

Perthshire (52)   1st (i) Grey (ii) Scarlet
              2nd (i) Grey (ii) Scarlet

1—Perth (1)	Perth (1)	Perth
2—(Breadalbane)	—	(The Perthshire Highland RV)
3— —	(Breadalbane) (2)	—
4— —	(Breadalbane) (2)	—
5—Blairgowrie	Blairgowrie (2)	—
6—Dunblane (1)	Dunblane (1)	—
7—Coupar Angus (1)	Coupar Angus (2)	—
8—Crieff (1)	Crieff (1)	—
9—Alyth (1)	Alyth (2)	—
10—Strathtay†	Strathtay (2)	—
11—Doune (1)	Doune (1)	—
12—Callander (1)	—	—
13—St Martins	St Martins (2)	—
14—Birnam	Birnam (2)	—
15—Auchterarder	Auchterarder (1)	—
16—Stanley	—	—
17—Comrie*	—	—

RIFLE VOLUNTEER CORPS

	1861	1872	1880
18—	—	Perth (1)	—
19—	—	Crieff (1)	—
20—	—	Pitlochrie (2)	—

† United, for drill and administrative purposes, with Breadalbane RVC (2).

4 VB R Highlanders (183) became 6 (Perthshire) Bn R Highlanders (The Black Watch) in 1908.
5 VB R Highlanders (184) became Highland Cyclist Bn.

Radnor (85)
1—Presteign   Presteign   —
2—Knighton   Knighton   —
3—New Radnor   New Radnor   —

By 1872, the three corps were united with 1 Adm Bn Herefordshire RV. Radnor recruits formed 1 Coy of 1 Bn Herefordshire Regt in 1908.

Renfrewshire (14) 1st (i) Grey (ii) Scarlet   2nd (i) Scarlet (ii) Blue
                   3rd (i) Scarlet (ii) Blue
1—Greenock (1)   Greenock (1)   (Renfrew and Bute)
2—   —   —   Paisley
3—Paisley (2)   Paisley (2)   Pollockshaws
4—Pollockshaws (3)   Pollockshaws (3)   —
5—Port Glasgow (1)   Port Glasgow (1)   —
6—Paisley (2)   Paisley (2)   —
7—Barrhead (3)   Barrhead (3)   —
8—Neilston (3)   Neilston (3)   —
9—Johnstone (2)   Johnstone (2)   —
10—Greenock (1)   (The Greenock Hldrs) (1)   —
11—Greenock (1)   —   —
12—   —   —   —
13—   —   —   —
14—Paisley (2)   Paisley (2)   —
15—Kilbarchan (2)   Kilbarchan (2)   —
16—Thornliebank (3)   Thornliebank (3)   —
17—Lochwinnoch (2)   Lochwinnoch (2)   —
18—   —   —   —
19—Hurlet (3)   Hurlet (3)   —
20—Renfrew (2)   Renfrew (2)   —
21—Barrhead (3)   Barrhead (3)   —
22—Gourock (1)   Gourock (1)   —
23—Cathcart (1)   Cathcart (3)   —
24—Paisley (2)   Paisley (2)   —
25—   —   Thornliebank (3)   —

1 VB Arg and Suth Highlanders (85) became 5 (Renfrewshire) Bn A and SH in 1908.
2 (86) and 3 (87) VB's Arg and Suth Highlanders became 6 (Renfrewshire) Bn A and SH in 1908.

1861	1872	1880
Ross-shire (38)	(i) Scarlet (ii) Blue	
1—Invergordon	Tain (1)	Dingwall
2—Dingwall	Dingwall (1)	—
3—Avoch	Avoch (1)	—
4—Knockbain	Knockbain (1)	—
5— —	Ullapool (1)	—
6— —	Invergordon (1)	—
7— —	Evanton (1)	—
8— —	Moy (1)	—
9— —	Gairloch (1)	—

Became 1 VB Seaforth Highlanders (152), and 4 (Ross Highland) Bn Seaforth Highlanders in 1908.

Roxburgh (34)	(i) Grey (ii) Grey	
1—Jedburgh	Jedburgh (1)	(The Border)
2—Kelso	Kelso (1)	—
3—Melrose	Melrose (1)	—
4—Hawick	Hawick (1)	—
5—Hawick*	—	—

By 1872, the Adm Bn was named The Border Corps and also contained 1st and 2nd Selkirkshire Corps.
The unit became 1st Roxburgh and Selkirk VRC (145) and amalgamated with 2 VB KOSB (185), to become 4 (The Border) Bn KOSB in 1908.

Selkirkshire (83)		
1—Galashiels	(see Roxburgh)	
2—Selkirk		

Shropshire (48)	1st (i) Scarlet (ii) White	
	2nd (i) Grey (ii) Black	
1—Shrewsbury (1)	Shrewsbury (1)	Shrewsbury
2—Market Drayton (2)	Market Drayton (2)	Shrewsbury
3—Whitchurch (2)	Whitchurch (2)	—
4—Bridgnorth (1)	Bridgnorth (1)	—
5—Condover (1)	Condover (1)	—
6—Much Wenlock (1)	Ironbridge (1)	—

RIFLE VOLUNTEER CORPS

1861	1872	1880
7—Wellington (2)	Wellington (2)	—
8—Hodnet (2)	Hodnet (2)	—
9—	—	—
10—Ludlow (1)	Ludlow (1)	—
11—Cleobury Mortimer (1)	Cleobury Mortimer (1)	—
12—Wem (2)	Wem (2)*	—
13—Ellesmere (2)	Ellesmere (2)	—
14—Shiffnall (1)	Shiffnall (1)	—
15—Oswestry (2)	Oswestry (2)	—
16—Munslow (1)	—	—
17—Shrewsbury (1)	Shrewsbury (1)	—
18— —	Newport (2)	—

Became 1 VB KSLI (174) and 2 VB KSLI (175) in 1881, and 4 Bn KSLI in 1908.

Somersetshire (45) 1st (i) Scarlet (ii) Black   2nd (i) Grey (ii) Black
               3rd (i) Grey (ii) Black

1—Bath (1)	Bath (1)	Bath
2—Bathwick (1)	Bathwick (1)	Taunton
3—Taunton (2)	Taunton (2)	Wells
4—Burnham (3)	Burnham (3)	—
5—Bridgwater (2)	Bridgwater (2)	—
6—Weston-super-Mare (3)	Weston-super-Mare (3)	—
7—Keynsham (1)	Keynsham (1)	—
8—Wellington (2)	Wellington (2)	—
9—Williton (2)	Williton (2)	—
10—Wells (3)	Wells (3)	—
11—Stogursey (2)	Nether Stowey (2)	—
12—Wiveliscombe (2)	Wiveliscombe (2)	—
13—Frome (3)	Frome (3)	—
14—Warleigh Manor (1)	Warleigh Manor (1)	—
15—Shepton Mallett (3)	Shepton Mallett (3)	—
16—Yeovil (2)	Yeovil (2)	—
17—Lyncombe (1)	Lyncombe (1)	—
18—Walcot (1)	Walcot (1)	—
19—Glastonbury (3)	Glastonbury (3)	—
20—Crewkerne (2)	Crewkerne (2)	—
21—Langport (2)	Langport (2)	—
22—Temple Cloud (1)	Kilmersdon (1)	—
23—Wincanton (3)	Castle Carey (3)	—
24—Somerton (3)	—	—
25—Baltonsborough (3)	Keinton (3)	—

RIFLE VOLUNTEER CORPS

	1861	1872	1880
26—	—	Bridgwater (2)	—
27—	—	Langford (3)	—

1 Corps became 1 VB Somerset LI (166) in 1880 and 4 Bn Somerset LI in 1908.

2 and 3 Corps became 2 VB Somerset LI (167) and 3 VB (168) in 1880 and 5 Bn Somerset LI in 1908.

Staffordshire (18) 1st (i) Scarlet (ii) White 2nd (i) Scarlet (ii) Blue
4th (i) Scarlet (ii) White 5th (i) Scarlet (ii) Blue
7th (i) Scarlet (ii) Blue

1861	1872	1880
1—Handsworth (3)	Handsworth (3)	Handsworth
2—Longton (1)	Longton (1)	Stoke
3—Hanley (1)	Hanley (1)	—
4—Walsall (5)	Walsall (5)	Walsall
5—Wolverhampton (4)	Wolverhampton (4)	Wolverhampton
6—Burslem (1)	Burslem (1)	—
7—Burton-on-Trent (2)	Burton-on-Trent (2)	Lichfield
8—Burton-on-Trent (2)	Burton-on-Trent (2)	—
9—Tunstall (1)	Tunstall (1)	—
10—Stoke (1)	Stoke (1)	—
11—Tipton (4)	Tipton (4)	—
12—Bilston (4)	Bilston (4)	—
13—Kidsgrove (1)	Kidsgrove (1)	—
14—Bloxwich (5)	Bloxwich (5)	—
15—Brierley (3)	Brierley (3)	—
16—Newcastle-under-Lyne (1)	Newcastle (1)	—
17—Seisdon (3)	Seisdon (3)	—
18—Kingswinford (3)	Kingswinford (3)	—
19—Tamworth (2)	Tamworth (2)	—
20—West Bromwich (3)	West Bromwich (3)	—
21—Rugeley (2)	Rugeley (2)	—
22—Brownhills (5)	Brownhills (5)	—
23—Wolverhampton (4)	Wolverhampton (4)	—
24—Lichfield (2)	Lichfield (2)	—
25—Stafford (2)	Stafford (2)	—
26—Willenhall (4)	Willenhall (4)	—
27—Potshull	Potshull (3)	—
28—Leek	Leek (1)	—
29—Sedgley (4)	Sedgley (4)	—
30—Tettenhall (4)	Tettenhall (4)	—
31—Smethwick	Smethwick (3)	—
32—Wolverhampton	Wolverhampton (4)	—
33—Cank (5)	Cank (5)	—

RIFLE VOLUNTEER CORPS

	1861	1872	1880
34	Wednesbury (5)	Wednesbury (5)	—
35	Kinver (3)	—	—
36	Hanley (1)	Hanley (1)	—
37	Cheadle (1)	Cheadle (1)	—
38	Eccleshall (1)	—	—
39	Burton-on-Trent	Burton-on-Trent (2)	—
40	Stone (1)	Stone (1)	—

The Adm Bns, numbered 1 to 5 were 1-Stoke, 2-Lichfield, 3-Handsworth, 4-Wolverhampton, 5-Walsall. They were re-numbered before 1880. 2 VB S Staffs (96) and 1 VB Staffs (94) became 5 Bn S Staffs in 1908.
3 VB S Staffs (97) became 6 Bn S Staffs in 1908.
1 VB N Staffs (95) became 5 Bn N Staffs in 1908.
2 VB N Staffs (98) became 6 Bn N Staffs in 1908.

Stirlingshire (23)	(i) Green (ii) Scarlet	
1—Stirling (1)	Stirling (1)	Stirling
2—Stirling (1)	Stirling (1)	—
3—Falkirk (1)	Falkirk (1)	—
4—Lennoxtown (1)	Lennoxtown (1)	—
5—Balfron (1)	Balfron (1)	—
6—Denny (1)	Bonnybridge (1)	—
7—Lennox Mill (1)	Lennox Mill (1)	—
8—Strathblane (1)	—	—
9—Bannockburn (1)	Bannockburn (1)	—
10—Stirling	—	—
11—Stirling (1)	Stirling (1)	—
12— —	Carron (1)	—
13— —	Kilsyth (1)	—
14— —	Alva†	—

† *United to 1 Adm Bn Clackmannan RV by 1872.*

4 (106) and 7 (217) VB Argyll and Sutherland Highlanders became 7 Bn Argyll and Sutherland Highlanders in 1908.

Suffolk (22)	1st (i) Green (ii) Black	
	6th (i) Grey (ii) Scarlet	
1—Ipswich (2)	Ipswich (2)	Ipswich
2—Framlingham	Framlingham (2)	—
3—Woodbridge (2)	Woodbridge (2)	—
4—Bungay (3)	Bungay (3)	—
5— —	Wickham Market (2)	—
6—Stowmarket (1)	Stowmarket (1)	Sudbury
7—Halesworth (3)	Halesworth (3)	—

H

1861	1872	1880
8—Saxmundham (2)	Saxmundham (2)	—
9—Aldeburgh (2)	Leiston (3)	—
10—Eye (1)	Eye (1)	—
11—Sudbury (1)	Sudbury (1)	—
12—Bosmere (1)	—	—
13—Bury St Edmunds (1)	Bury St Edmunds (1)	—
14—Beccles (3)	Beccles (3)	—
15—Wrentham	—	—
16—Hadleigh (1)	Hadleigh (1)	—
17—Lowestoft (3)	Lowestoft (3)	—
18—Wickham Brooks	Newmarket (1)	—

The 3 Adm Bns of 1872 were 1—Sudbury, 2—Woodbridge, 3—Lowestoft. By 1880 there were only 2 Bns, Nos 1 and 6 as shown above. They became 1 (104) and 2 (105) VB's Suffolk Regt shortly afterwards, and became 4 and 5 Bns Suffolk Regt in 1908.

Recruits also joined the Essex and Suffolk Cyclist Bn, forming 4 Suffolk coys.

Surrey (4)   1st (i) Green (ii) Scarlet   2nd (i) Green (ii) Scarlet
             3rd (i) Green (ii) Rifle Green 4th (i) Green (ii) Scarlet
             5th (i) Green (ii) Scarlet   6th (i) Scarlet (ii) Blue
             7th (i) Green (ii) Scarlet   8th (i) Green (ii) Scarlet
1—(South London)   (South London)   (South London)
2—Croydon (1)   Croydon   Croydon
3—   —   Clapham
4—Brixton (1)   Brixton (1)   Dorking
5—Reigate (3)   Reigate (3)   Kingston-on-Thames
6—Esher (2)   Esher (2)   Rotherhithe
7—Southwark   Southwark   Southwark
8—Epsom (1)   Carshalton (1)   Kennington

RIFLE VOLUNTEER CORPS

1861	1872	1880
9—Richmond (2)	Richmond (2)	—
10—Bermondsey	Bermondsey (4)	—
11—Wimbledon (2)	Wimbledon (1)	—
12—Kingston-on-Thames	Kingston-on-Thames (2)	—
13—Guildford (3)	Guildford (3)	—
14—Dorking (3)	Dorking (3)	—
15—Chertsey (2)	Chertsey (2)	—
16—Egham (2)	—	—
17—Godstone (3)	Godstone (3)	—
18—Farnham (3)	Farnham (3)	—
19—Lambeth	Lambeth	—
20—Lower Norwood (1)	—	—
21—Battersea	—	—
22—No title	Albury (3)	—
23— —	Rotherhithe (4)	—
24— —	Guildford (3)	—
25— —	Epsom (1)	—

1 Surrey RVC (54) became 1 VB E Surrey in 1881, 21 (County of London) Bn (1st Surrey Rifles) in 1908 and were converted to RE in 1937.

6 Surrey RVC became 3 VB Queen's R W Surrey (59) in 1881, 22 (County of London) Bn (The Queen's) in 1908 and 6 (Bermondsey) Bn, The Queen's Royal Regt in 1937.

7 and 26 Corps were amalgamated in 1880, became 4 VB E Surrey (60) in 1881, 23 (County of London) Bn in 1908 and 7 (23rd London) Bn E Surrey Regt in 1937.

8 Corps became 4 VB Queen's R W Surrey (61) in 1881, 24 (County of London) Bn in 1908 and 7 (Southwark) Bn Queen's Royal Regt in 1937.

1 VB Queen's R W Surrey (55) became 4 Bn Queen's R W Surrey in 1908.

2 VB Queen's R W Surrey (57) became 5 Bn Queen's R W Surrey in 1908.

2 VB E Surrey (56) became 5 Bn E Surrey in 1908.

3 VB E Surrey (58) became 6 Bn E Surrey in 1908.

	1st (i) Scarlet (ii) Blue	
Sussex (1)	2nd (i) Scarlet (ii) Blue	
1—Brighton (3)	Brighton	Brighton
2—Cuckfield (3)	Cuckfield (2)	Worthing
3— —	—	—
4—Lewes (3)	Lewes‡	—
5—East Grinstead (3)	East Grinstead (2)	—
6—Petworth (2)	Petworth (2)	—
7—Horsham	Horsham (2)	—
8—Storrington (2)	Storrington (2)	—

	1861	1872	1880
9—	Arundel (1)	Arundel (1)	—
10—	Chichester (1)	Chichester (1)	—
11—	Worthing (1)	Worthing (1)	—
12—	Westbourne (1)	Westbourne (1)	—
13—	Hurstpierpoint (2)	Hurstpierpoint (2)	—
14—	Crawley (2)	—	—
15—	Bognor (1)	—	—
16—	Battle (3)	Battle ‡	—
17—	Etchingham†	Etchingham‡	—
18—	Henfield	Henfield (2)	—
19—	Eastbourne	—	—
20—	Billingshurst*	Uckfield‡	—

† *United for admin and drill with 37 Kent RVC.*
‡ *United with 1 Adm Bn Cinque Ports RV.*

1 VB R Sussex (74) was converted to 2 Home Counties Bde RFA, and Homes Counties Divisional HQ Coy ASC in 1908.
2 VB R Sussex (75) became 4 Bn R Sussex in 1908.
1st Cinque Ports VRC (146) became 5 (Cinque Ports) Bn R Sussex in 1908.

Sutherland (54)	(i) Scarlet (ii) Yellow	
1—Golspie	Golspie (1)	(The Sutherland Highland)
2— —	Dornoch (1)	—
3—No title	Brora (1)	—
4— —	Roggart (1)	—
5— —	Bonar Bridge (1)	—

By 1872, the Admin Bn—The Sutherland Highland RV—consisted of the five corps listed, plus 1st Orkney, 1st, 2nd, 3rd and 4th Caithness RVC. In 1908 the Unit became 5 (The Sutherland and Caithness Highland (Bn Seaforth Highlanders. Its precedence in the Volunteer Battalion list was 186.

Tower Hamlets (89)	1st (i) Scarlet (ii) Blue 2nd (i) Grey (ii) Scarlet	
1— —	(The Tower Hamlets RV Bde), Hoxton	(The Tower Hamlets RVB)
2—Hackney	—	Commercial Street, E
3—Spitalfields	Bethnal Green (1)	—
4—St Leonards, Shoreditch	—	—
5—Dalston	—	—

RIFLE VOLUNTEER CORPS

1861	1872	1880
6—Dalston	(North East London), Hoxton	—
7—Mile End	Mile End (1)	—
8—West India Docks, Poplar†	—	—
9—London Dock House	—	—
10—Goodman's Fields	Bethnal Green (1)	—

† United with 2 Tower Hamlets RVC as an Admin Bn by 1861.

By 1872 Nos 3, 7 and 10 Corps comprised No 1 Admin Bn with HQ in Bethnal Green.

2 and 4 Tower Hamlets VRC were amalgamated in 1868 to become 1 TH RV Bde, in 1874 6 (North East London) TH RVC joined them. The Bde was attached to the Rifle Bde in 1881, became 4 VB The London Regt in 1903, 4 Bn The London Regt (R Fusiliers) in 1908 and were converted to RE in 1936.

2 Tower Hamlets VRC (218) became 17 Bn London Regt in 1908. (See notes on this Bn under Middlesex.)

Warwickshire (41)	1st (i) Green (ii) Scarlet 2nd (i) Scarlet (ii) Blue	
1—(Birmingham)	(Birmingham)	(Birmingham)
2—Coventry (2)	Coventry (1)	Coventry
3— —	Rugby (1)	—
4—Rugby (2)	Warwick (1)	—
5—Warwick (2)	Stratford-upon-Avon (1)	—
6— —	—	—
7—Stratford-upon-Avon (2)	—	—
8—Coventry	Nuneaton (1)	—
9—Coventry (2)	Saltley (1)	—
10—Nuneaton	Leamington (1)	—

By 1872 the Admin Bn—No 1—was in Coventry.

1 Corps became 1 VB R Warks (156) in 1880 and 5 and 6 Bns R Warks in 1908.

2 Corps became 2 VB R Warks (157) in 1880 and 7 Bn R Warks in 1908.

8 Bn R Warks was formed in 1908, recruits coming from Aston district.

Westmorland (78)	(i) Scarlet (ii) Green	
1—Lunesdale (1)	Lunesdale (1)	Kendal
2— —	—	—
3—Kendal (1)	Kendal (1)	—

1861	1872	1880
4—Windermere (1)	Windermere (1)	—
5—Ambleside (1)	Ambleside (1)	—
6—No title (1)	Grasmere (1)	—

By 1872 the 1st Admin Bn was in Kendal.
Became 1 VB R Lancs Regt, 2 (Westmorland) VB Border Regt and 4 (Cumberland and Westmorland) Bn Border Regt in 1908.

Wigtown (65)

1861	1872	1880
1—Wigton (1)	Wigton (1)	—
2—Stranraer (1)	Stranraer (1)	—
3—Newton Stewart (1)	Newton Stewart (1)	—
4—Whithorn (1)	Whithorn (1)	—
5—Dunmore	—	—

By 1872 the 4 Wigtown Corps and the 5 Kirkcudbright Corps were combined into 1st (Newton Stewart) Admin Bn.

Wiltshire (9)

1st (i) Green (ii) Black
2nd (i) Green (ii) Black

1861	1872	1880
1—Salisbury	Salisbury (1)	Warminster
2—Trowbridge	Trowbridge (1)	Chippenham
3—Malmesbury	Malmesbury (2)	—
4—Chippenham	Chippenham (2)	—
5—Devizes	Devizes (2)	—
6—Maiden Bradley	Maiden Bradley (1)	—
7—Market Lavington	Market Lavington (2)	—
8—Mere	Mere (1)	—
9—Bradford	Bradford (1)	—
10—Warminster	Warminster (1)	—
11—Swindon	Swindon (2)	—
12—Melksham	Melksham (2)	—
13—Westbury	Westbury (1)	—
14—Wilton	Wilton (1)	—
15—Wooton Bassett	Wooton Bassett	—
16—Old Swindon	Old Swindon (2)	—
17—Marlborough	Marlborough (2)	—
18—Highworth	Highworth (2)	—

1st Corps retained its title, 2nd Corps became 2 VB Wiltshire Regt and both were re-named 4 Bn Wiltshire Regt in 1908.

RIFLE VOLUNTEER CORPS

1861	1872	1880
Worcestershire (39)	1st (i) Green (ii) Green	
	2nd (i) Green (ii) Green	
1—Wolverley (1)	Wolverley (1)	Hagley
2—Tenbury (1)	Tenbury (1)	Worcester
3—Kidderminster (1)	Kidderminster (1)	—
4—Kidderminster (1)	Kidderminster (1)	—
5—Bewdley (1)	Bewdley (1)	—
6—Halesowen (1)	Halesowen (1)	—
7—Dudley (1)	Dudley (1)	—
8—Stourport (1)	Stourport (1)	—
9—Stourbridge (1)	Stourbridge (1)	—
10—Pershore (2)	Pershore (2)	—
11—Great Malvern (2)	Great Malvern (2)	—
12—Evesham (2)	Evesham (2)	—
13—Worcester (2)	Worcester (2)	—
14—Worcester (2)	Worcester (2)	—
15—Ombersley (2)	—	—
16—Oldbury (1)	Oldbury (1)	—
17—Redditch (2)	Redditch (2)	—
18—Droitwich (2)	Droitwich (2)	—
19—Upton-on-Severn	Upton-on-Severn (2)	—
20—Kidderminster (1)	Kidderminster (1)	—
21— —	Bromsgrove (2)	—

1 Corps became 1 VB Worcestershire Regt (153) and 7 Bn Worcestershire Regt in 1908.

2 Corps became 2 VB Worcestershire Regt (154) and 8 Bn Worcestershire Regt in 1908.

Yorkshire	1st (i) Scarlet (ii) Yellow	
(East Riding) (50)	2nd (i) Scarlet (ii) Buff	
1—Hull	Hull	Hull
2— —	—	Beverley
3—Howden (2)	Howden (1)	—
4— —	—	—
5—Bridlington (2)	Bridlington (1)	—
6—Beverley (2)	Beverley (1)	—
7— —	—	—
8—Driffield	Driffield (1)	—
9—Market Weighton	Market Weighton (1)	—
10—Hedon	Hedon (1)	—
11— —	Pocklington (1)	—

1 Corps became 1 VB E Yorks Regt (179) in 1881 and 4 Bn E Yorks Regt in 1908.

2 Corps became 2 VB E Yorks Regt (180) in 1881, joined 2 VB (202) Yorks Regt to become 5 Bn Yorks Regt in 1908.
Recruits also joined 5 (Cyclist) Bn E Yorks Regt from 1908.

1861	1872	1880
Yorkshire	1st (i) Scarlet (ii) Green	
(North Riding) (67)	2nd (i) Grey (ii) Scarlet	
1—Malton (2)	Malton (2)	Richmond
2—Swaledale (1)	—	Scarborough
3—Hovingham (2)	Hovingham (2)	—
4—Leyburn (1)	Leyburn (1)	—
5—Forcett (1)	Gilling (1)	—
6—Scarborough (2)	Scarborough (2)	—
7—Startforth (1)	—	—
8—Bedale (1)	Bedale (1)	—
9—Stokesley (1)	Stokesley (1)	—
10—Helmsley (2)	Helmsley (2)	—
11—Masham (1)	—	—
12—Carperby (1)	Thornton Rust (1)	—
13—Thirsk (2)	—	—
14—Catterick (1)	Catterick (1)	—
15—Richmond (1)	Richmond (1)	—
16—Pickering Lythe (2)	Pickering Lythe (2)	—
17—Pickering by the East (2)	—	—
18—Skelton (1)	Skelton (1)	—
19—Northallerton (1)	Northallerton (1)	—
20—Whitby*	Gisbro' (1)	—

1 VB Yorks Regt (201) became 4 Bn Yorks Regt in 1908.
2 VB Yorks Regt (202) amalgamated with 2 VB E Yorks Regt (180), and formed 5 Bn Yorks Regt in 1908.

Yorkshire (West Riding) (3)
    1st (i) Scarlet (ii) Blue   2nd (i) Scarlet (ii) White
    3rd (i) Scarlet (ii) D Green   4th (i) Scarlet (ii) Blue
    5th (i) Scarlet (ii) Blue   6th (i) Scarlet (ii) Sky Blue
    7th (i) Grey (ii) Grey   8th (i) Scarlet (ii) Green
    12th (i) Scarlet (ii) Buff

1—York (1)	York (1)	York
2—(Hallamshire)	(Hallamshire)	(Hallamshire)
3—Bradford	Bradford†	Bradford
4—Halifax	Halifax	Halifax
5—Wakefield (3)	Wakefield (3)	Wakefield
6—Huddersfield	(The Huddersfield) (5)	Huddersfield

7—Leeds	Leeds	(Leeds)
8—	—	Doncaster
9—	—	—
10—	—	—
11—	—	—
12—Skipton (2)	Skipton (2)	Skipton
13—	—	—
14—	—	—
15—North Craven (2)	Settle (2)	—
16—Harrogate (1)	Harrogate (1)	—
17—Knaresborough (1)	Knaresborough (1)	—
18—Pontefract (4)	Pontefract (4)	—
19—Rotherham (4)	Rotherham (4)	—
20—Doncaster (4)	Doncaster (4)	—
21—Doncaster (4)	Doncaster (4)	—
22—	—	—
23—Burley (2)	Burley (2)	—
24—	—	—
25—Guiseley (2)	Guiseley (2)	—
26—Ingleton (2)	Ingleton (2)	—
27—Ripon (1)	Ripon (1)	—
28—Goole (3)	Goole (3)	—
29—Dewsbury (3)	Dewsbury (3)	—
30—Birstal (3)	Birstal (3)	—
31—Tadcaster (1)	Tadcaster (1)	—
32—Holmfirth	Holmfirth (5)	—
33—	—	—
34—Saddleworth	(Saddleworth)	—
35—Keighley	(Airdale) (2)	—
36—Rotherham	Rotherham (4)	—
37—Barnsley	Barnsley (4)	—
38—Selby (3)	Selby (3)	—
39—	Saltaire‡	—
40—	Wath-upon-Dearne (4)	—
41—	Mirfield (5)	—
42—	Haworth (2)	—
43—	Batley (3)	—
44—	Meltham (5)	—

† *Attached to 39 W Riding of Yorkshire RVC.*
‡ *Attached to 3 W Riding of Yorkshire RVC.*

1 Corps became 1 VB W Yorks Regt (129) in 1881 and 5 Bn W Yorks Regt in 1908.

3 Corps became 2 VB W Yorks Regt (131) in 1881 and 6 Bn W Yorks Regt in 1908.

7 Corps became 3 VB W Yorks Regt (135) and 7 and 8 Bns W Yorks Regt in 1908.

4 Corps became 1 VB W Riding Regt (132) and 4 Bn W Riding Regt in 1908.
7 Coys of 2 VB W Riding Regt (134) formed 5 Bn W Riding Regt in 1908.
3 VB W Riding Regt (137) became 6 Bn W Riding Regt in 1908.
3 Coys of 2 VB W Riding Regt (134) formed 7 Bn W Riding Regt in 1908.
Part of 1 VB Yorks LI (132) formed 4 Bn Yorks LI in 1908.
Part of 1 VB Yorks LI (133) formed 5 Bn Yorks LI in 1908.
2 Corps became 1 Hallamshire VB York and Lancaster Regt (130) and 4 (Hallamshire) Bn York and Lancaster Regt in 1908.
2 VB York and Lancaster Regt (136) became 5 Bn York and Lancaster Regt in 1908.

Appendix 1

OFFICERS' RANK DESIGNATIONS FOR RIFLE REGIMENTS CURRENT IN 1860

Rank	Collar badges	Lace
Colonel Lieut-Colonel Major	crown and star crown star	Collar laced all around with black lace with figured braiding within the lace. Sleeve ornament of lace and figured braiding, 11in deep
Captain	crown and star	Collar laced all around the top with black lace, with figured braiding below the lace. Sleeve ornament a knot of square cord with figured braiding, 8in deep
Lieutenant Ensign	crown star	Collar laced all around the top with black lace, with edging of plain braid. Sleeve ornament a knot of square cord and braid, 7in deep

All collar badges in silk embroidery

Appendix 2

VOLUNTEER FORCE REGULATIONS—19 JANUARY 1861

Extracts relating to uniform, clothing and accoutrements.

95 Every volunteer corps is allowed to choose its own uniform and accoutrements, subject to the approval of the lord lieutenant of the county, and provided no gold lace is introduced.

96 It is desirable that a uniform colour should be chosen for the clothing of corps of each arm within the same county. This is most important in the case of corps which are likely to be united together in administrative regiments, brigades or battalions.

97 The distinctions in uniform and appointments which are prescribed in the regular service and militia to denote the rank of the wearer should be observed strictly by volunteers of the various grades, as far as they are applicable to the volunteer force. In this respect the dress regulations for the Army are to be taken as a guide.

98 A list of additional kit which members should purchase.

99 The pouches should be capable of containing sixty rounds of ammunition, and should be so fixed as not to interfere with the arrangement of the knapsack.

100 Commissioned officers and sergeants alone are permitted to wear side arms when off duty, and then only the authorised weapons of their respective ranks.

101 Neither standards nor colours are to be carried by corps on parade, as the volunteer force is composed of arms to which their use is not appropriate.

Appendix 3

OFFICERS' RANK DESIGNATIONS—VOLUNTEERS, 1881

Only chaplains will continue to wear rank badges on collars, all other officers will wear badges on shoulder straps on tunics, stabling jackets and shell jackets. Shoulder straps to be of the same material as the garment. Scarlet, blue and green straps to be edged with half-inch black mohair braid, except at the base, with black netted button at the top. Grey straps to be edged with grey braid, with grey netted button at the top.

Where knots were used on shoulders, the following to apply:
Corps clothed in scarlet—Universal pattern, in silver.
Corps clothed in grey with silver on sleeves—Universal pattern, in silver.
Corps clothed in grey with cord other than silver on sleeves—Universal pattern, colour and material to be as cord on sleeves.
Corps clothed in green—Black chain gimp.

Rank badges:

Colonel	Crown and two stars below	Badges to be in gold embroidery on straps of tunic, stable jacket and shell jacket.
Lieut-Colonel	crown and one star below	
Major	crown	Badges to be in silver embroidery on straps of frock, patrol jacket, cloak and greatcoat
Captain	two stars	
Lieutenant	one star	
Supernumerary Lieutenant	none	

Appendix 4

SOME BOOKS FOR THE COLLECTOR

Acklom, Capt J. E. *Origin of the present Volunteer movement* (1862)
Anon. *Digest of the Law relating to Volunteer Corps* (1803)
Anon. *Drill and Rifle Instruction for the Corps of Rifle Volunteers* (1859)
Bannantyne, Maj M. *Our Military Forces and Reserves* (1867)
Baxter, R. D. *The Volunteer movement, its progress and wants* (1860)
Berry, R. P. *A history of the formation and development of the Volunteer Infantry, illustrated by local records of Huddersfield and its vicinity, from 1794–1874* (1903)
Boynton, L. *The Elizabethan Militia, 1558–1638* (1967)
Cave, Col T. S. *History of the 1st V. B. Hampshire Regiment, 1859–1889* (1905)
Coles, Capt C. *Our National Defences* (1861)
Collyer, J. N. and Pocock, J. J. *An historical record of the Light Horse Volunteers of London and Westminster* (1843)
Corner, W. *The story of the 34th Company (Middlesex) Imperial Yeomanry* (1902)
Cousins, G. *The Defenders; a history of the British Volunteer* (1968)
Davenport, Lt-Col. *The Light Horse Drill, describing the several evolutions, for the Volunteer Corps* (c1800)
Edmeades, Lt-Col J. F. *Some historical records of the West Kent (Queen's Own) Yeomanry, 1794–1909* (1909)
Erskine, T. *Opinion on the Volunteer Acts* (1804)
Evans, Lt E. T. *Record of the 3rd Middlesex Rifle Volunteers and of the Corps which formed the 2/6 Middlesex Administrative Battalions, 1794–1884* (1885)
Evans, Capt H. C. *Records of the 4th V.B. Manchester Regiment, 1859–1900* (1900)
Evans, Capt H. D. *Development of the reserve forces* (1875)
Fellows, Capt G. *History of the South Notts Yeomanry Cavalry, 1794–1894* (1895)
Fisher, W. G. *History of the Somerset Yeomanry, Volunteer and Territorial Units, 1745–1923* (1924)
Freeman, B. *The Yeomanry of Devon, 1794–1927* (1927)
Gibney, R. W. *The history of the 1st Bn Wiltshire Volunteers, 1861–85* (1888)

APPENDIX

Grierson, Maj-Gen J. M. *Records of the Scottish Volunteer Force, 1859–1908* (1909)

Hadrill, Sgt-Major, *The Volunteer Rifleman's Manual* (c1860)

Hay, Col G. J. *An epitomised history of the Militia (the Constitutional Force), with service of Militia Units existing on 31st October, 1905* (1905)

Hayhurst, T. H. *A history and some records of the Volunteer movement in Bury, Heywood, Rossendale and Ramsbottom from 1642* (1877)

Holden, Capt R. *Historical Record of the Third and Fourth Battalions of the Worcestershire Regiment, 1778–1886* (1887)

Jewitt, L. *Rifles and Volunteer Rifle Companies, their constitution, arms, drill, laws and uniform* (c1860)

Kinloch, Capt J. *Proposal for the defence of the Country by means of a Volunteer Force* (1852)

Maurice, Maj-Gen Sir F. *The history of the London Rifle Brigade, 1859–1919*, 2 vols (1921)

Milne, Capt B. A. *Historical record of the 1st Cornwall (Duke of Cornwall's) Artillery Volunteers* (1885)

Montefiore, Capt C. S. *A history of the Volunteer Forces, from the earliest times to 1860* (1908)

Napier, Sir C. J. *A letter on the defence of England by Volunteers and Militia* (1852)

Owen, H. J. *Merioneth Volunteers and Local Militia during the Napoleonic Wars, 1795–1816* (1934)

Pease, H. *The History of The Northumberland (Hussars) Yeomanry, 1819–1919* (1924)

Raikes, Capt G. A. *The History of the Honourable Artillery Company*, 2 vols (1878–9)

Richards, W. *His Majesty's Territorial Army*, 4 vols (1910–11)

Stephen, W. *History of the Queen's City of Edinburgh Rifle Volunteer Brigade* (1881)

Stonham, Maj C. and Freeman, B. *Historical Records of the Middlesex Yeomanry, 1797–1927* (1930)

Swann, Maj-Gen J. C. *The Citizen Soldiers of Buckinghamshire, 1795–1926* (1930)

Tamplin, J. M. *The Lambeth and Southwark Volunteers, 1860–1960* (1965)

Terry, Capt A. *Historical record of the 5th Administrative Battalion Cheshire Rifle Volunteers* (1879)

Thompson, Col A. *A history of the Fife Light Horse, 1860–1892* (1892)

Thoyts, Miss E. E. *History of the Royal Berkshire Militia* (1897)

Townshend, G. *A plan of the discipline composed for the use of the Militia of the County of Norfolk* (1759)

Walter, Maj J. *The Volunteer Force: History and Manual* (1881)

Wheeler-Holehan, Capt A. V. and Wyatt, Capt G. M. *The Rangers historical record, 1859–1919* (c1921)

White, A. S. *A bibliography of Regimental Histories of the British Army* (1965)

Wild, Col E. T. *The Tower Hamlets Rifle Volunteer Brigade (1st Tower Hamlets Rifle Volunteers), from May 1903 the 4th V.B. Royal Fusiliers* (1910)

Young, Sir W. *Instructions for the Armed Yeomanry* (c1803)

Bibliography

Blackmore, H. L. *British military firearms 1650–1850* (1961)
Bowling, A. H. *British infantry regiments 1660–1914* (1970)
Dress regulations for the Army—1900. David & Charles reprint (1970)
Field Service Pocket book, 1914. David & Charles reprint (1972)
Gaylor, J. *Military badge collecting* (1971)
Guns Review. Monthly magazine. Article on the British Volunteer Rifle, 1850–65 (December 1968); article on Whitworth rifles (March 1969); article on Enfield Volunteer Rifles (May, June 1970)
Ripley, H. *Buttons of the British Army 1855–1970* (1971)
Roads, C. H. *The British soldier's firearm 1850–64* (1964)
Wagner, E. *Cut and thrust weapons* (1967)
Wallis & Wallis priced illustrated auction catalogues
Walter, J. and Hughes, G. *A primer of world bayonets* (two pamphlets, 1969)
Wilkinson Latham, J. *British military swords from 1800 to the present day* (1966)
———. *British military bayonets from 1700 to 1945* (1967)
———. *Regulation military swords* (1970)
———. *British cut and thrust weapons* (1972)

Index

Page numbers in italics denote illustrations

Accoutrements, 56, 58
Administrative battalions, 20; Hampshire RV, 20
Alt, Lt-Col W., 70
Armed Associations, 14
Army List, 31–2
Army Museums Ogilby Trust, 33
Army Postal Corps, 24
Army re-organisation of 1881, 23
Arquebus, 12
Artillery, 12
Assize of Arms of 1181, 11

Badges: Artists, *90*; Artists Rifles, *90*; Berkshire IVB, *107*; Berkshire RVC, *89*; Devon Yeomanry, *55*; East Riding Yeomanry, *56*; Hampshire 2VB, *90*; KOSB 2VB, *89*; London Regt 13Bn, *55*; London Rifle Brigade, *108*; London Scottish, *90*; Loyal Suffolk Hussars, *108*; Middlesex 24 RVC, *89*; Officer's rank 1860, 135; Officer's rank 1881, 137; proficiency, 43–4; Rangers, *90*; Royal Gloucestershire Hussars, *90*; Suffolk 2VB, *107*; Tyneside Scottish, *90*
Bailey, D., 74
Bayonets, 66–7; markings, 67–8
Bisley rangers, 24
Boer War, 24, 60–2
Bows, 11–12

Bristol Volunteers medal, 63–4, *72*
Busby, 51
Buttons, 43, 58–9; makers, 59

Cadet Corps, 21; battle honours, 23
Camps for annual training, 21
Carmen, WY, 50
Catholics, not eligible for militia, 13
Charles I, 13
City Imperial Volunteers, 23, 60–1; lapel badge, *24*
Civil War, 13
Commissions of Muster, 12
Consolidated battalions, 20
Crimean War, 60
Cromwell, 13
Cuff lace patterns, 40

Dress, importance of to volunteers, 19
Drill and Rifle Instructions for Volunteer Rifle Corps, 21; title page, *22*
Duke of Cumberland's Sharpshooters, 16, 26

Edinburgh Rifle Volunteers, 20
Egerton, T., 30
Egypt, 24
Enfield 0·577 rifle, 19, 73–4

INDEX

Field Telegraph Corps, 24
Firearms Act, 69
Firearms, first use of, 12
French invasion, 11, 14

Gaylor, J., 56
Glengarry, 51

Hampshire RV, first meeting, 19; precedence, 20; uniform, 45
Hart's Army List, 31
Helmet, Home Service Pattern, 52, *18*
Home Guard, 14
Honourable Artillery Company, 16, 26
Hyde Park reviews, 24

Imperial Yeomanry, 24; Long Service medal, 61; uniform, *17*

Kerr rifles, 74
Khaki, adoption of, 49
King's Regulations, of 1837, 26
KRRC, 1st Cadet Bn, 21

Levy *en masse* Act of 1803, 15
Liverpool Drill Club, 16
Local Militia Act of 1808, 15
London bands, 13
London Rifle Brigade, 21, 51; Leaflet, *28*
Lords Lieutenant, 12, 14, 16, 39
Loyal Volunteers of London and Environs, 29

Machine guns, 70
Martin, E. J., 32
Martini rifles, 75
Maxim guns, 70

Medals, Bideford award 61–2, *62*; Imperial Yeomanry Long Service, 63; King's South Africa, 60; Queen's South Africa, 61; Royal Bristol Volunteers, 63–4, *72*; Territorial Decoration, 62; Territorial Efficiency Medal, 63; unofficial awards, 64–5; Volunteer Officer's Decoration, 63
Middlesex, 17 RVC, 51; 22 RVC, 70; 24 RVC, 24; 26 RVC, 70; 29 RVC, 32; 49 RVC, 24
Militia, 12–13, 23; Act of 1757, 13; augmentation in 1794, 14
Monk, Gen, 13
Moore, Capt Denis, 16

Napier, Sir Charles, 21
Napoleon III, 15
Nordenfelt guns, 70
Norman conquest, 11

Orsini, 15

Parker Field & Sons, 67
Peel, Gen, 16
Plumes, 51–2
Post Office Volunteers, 24
Precedence, order of, 19
Proficiency badges, 45
Punch, 41–2; cartoon, *42*

Queen's Own Cameron Highlanders, medals, 61
Queen's Regulations, 31
Queen's Rifle Volunteers, 23

Raikes, G. A., 31

INDEX

Rank badges, officers', 40–1, 135, 137; OR's, 41
Reilly, 67
Rifle Volunteer Corps, 1st, 16
Rifle Volunteer Counties:
　Aberdeenshire, 79
　Anglesey, 79
　Argyll, 80
　Ayrshire, 80
　Banffshire, 81
　Bedfordshire, 81
　Berkshire, 81
　Berwickshire, 82
　Berwick-on-Tweed, 82
　Brecknockshire, 82
　Buckinghamshire, 83
　Buteshire, 83
　Caithness, 83
　Cambridgeshire, 83
　Cardiganshire, 84
　Carmarthenshire, 84
　Carnarvonshire, 84
　Cheshire, 84–5
　Cinque Ports, 85–6
　Clackmannon, 86
　Cornwall, 86
　Cumberland, 87
　Denbighshire, 87
　Derbyshire, 88
　Devonshire, 88–9
　Dorsetshire, 91
　Dumbarton, 91–2
　Dumfries, 92
　Durham, 92–3
　Edinburgh (City), 93
　Essex, 94–5
　Fifeshire, 95
　Flintshire, 95
　Forfar, 95–6
　Glamorgan, 96
　Gloucestershire, 96–7
　Haddington, 97
　Hampshire, 97–8
　Haverfordwest, 98
　Herefordshire, 98
　Hertfordshire, 99
　Huntingdon, 99
　Inverness-shire, 99
　Isle of Wight, 100
　Kent, 100–1
　Kincardineshire, 101
　Kirkudbrightshire, 102
　Lanarkshire, 102–5
　Lancashire, 105–10
　Leicestershire, 110–11
　Lincolnshire, 111
　Linlithgowshire, 111
　London, 112
　Merionethshire, 112
　Middlesex, 112–16
　Midlothian, 116
　Monmouthshire, 116
　Montgomeryshire, 117
　Nairn, 117
　Newcastle-upon-Tyne, 117
　Norfolk, 117–18
　Northamptonshire, 118
　Northumberland, 118–19
　Nottinghamshire, 119
　Orkney, 119
　Oxfordshire, 119–20
　Peeblesshire, 120
　Pembrokeshire, 120
　Perthshire, 120–1
　Radnor, 121
　Renfrewshire, 121–2
　Ross-shire, 122
　Roxburgh, 122

INDEX

Rifle Volunteer Counties—cont.
  Shropshire, 122–3
  Somersetshire, 123–4
  Staffordshire, 124–5
  Stirlingshire, 125
  Suffolk, 125–6
  Surrey, 126–7
  Sussex, 127–8
  Sutherland, 128
  Tower Hamlets, 128–9
  Warwickshire, 129
  Westmorland, 129–30
  Wigtown, 130
  Wiltshire, 130
  Worcestershire, 131
  Yorkshire, 131–2
  Yorkshire (North Riding), 132
  Yorkshire (West Riding), 132–4
Rowlandson, 29
Royal Commission of 1862, 21
Royal Masonic School Cadets, 21
Royal Victoria Rifle Regiment, 16, 26, 41

Shako, 51
Snider rifle, 75
Society for Army Historical Research, 32
Spanish Armada, 12
Special Reserve, 25
Standing Army, 13
Statute of Winchester, 11

Swords, 70, 73

Territorial and Reserve Forces Act of 1907, 25
Territorial Army, 25
Territorial Force, 25
Territorial Force Associations, 25
Tokens, 64–5
Train bands, 13
Turner rifles, 74

Uniforms, care, 34; cost, 20; importance to volunteers, 19

Volunteer Active Service Coys, 25
Volunteer Act of 1794, 14
Volunteer, magazines, 32; service coys, 60–1
Volunteers, numbers in 1812, 15; numbers in 1873, 23; numbers in 1886, 23

Wallis & Wallis auctions, 50
Welsh Regt, 3VB, 23
Westley Richards rifle, 75
White, A. S., 32
Whitworth rifles, 74
Wimbledon ranges, 24
Windsor Park reviews, 24

Yeomanry crests, *57*